Lecture Notes in Electrical Engineering

Volume 183

For further volumes:
http://www.springer.com/series/7818

Dhanasekharan Natarajan

A Practical Design of Lumped, Semi-Lumped and Microwave Cavity Filters

 Springer

Dhanasekharan Natarajan
RV College of Engineering
Bangalore
India

ISSN 1876-1100   ISSN 1876-1119   (electronic)
ISBN 978-3-642-42861-6   ISBN 978-3-642-32861-9   (eBook)
DOI 10.1007/978-3-642-32861-9
Springer Heidelberg New York Dordrecht London

Printed on acid-free paper

Springer is part of Springer Science+Business Media (www.springer.com)

Dedicated to OM SAI

Preface

In the design of electronic equipments, additional frequency components outside the frequency band of interest are generated for various reasons such as interference and nonlinearity. The additional frequency components are unwanted band of frequencies as they affect the satisfactory functioning of the equipments. Filter circuits are inserted between appropriate electronic modules in the equipments to pass the required band of frequencies with minimum attenuation and to reject the unwanted band of frequencies with maximum attenuation. The frequency band of filters varies with the design requirements of equipments. Considering size, frequency band and other electrical characteristics of filters, a host of technologies is available for designing the filters.

Commercial electromagnetic field simulator is popular for designing RF (Radio Frequency) filters. Vast microwave literature on filter design technologies is available and they are also used for designing RF filters. The application of microwave literature for designing lumped/semi-lumped filters and combline/iris-coupled microwave cavity filters is presented. The design of the filters is based on the established design procedures from the Microwave Filter Design Book [1] and practical industrial experience. The design equations and graphs available in the various chapters of the Microwave Filter Design Book are organised in the form of tutorials and the design of the filters is illustrated with examples to enthuse confidence among students and young entrepreneurs to become RF filter designers. The tutorial presentations could be computerised to save design time of RF filters. Physical understanding of the terms and characteristics of RF filters is emphasised in the book. Mathematical treatment is limited to the level that supports physical understanding. The book complements engineering textbooks on RF components and provides valuable support for the project assignments of students.

Chapter 1 explains the need for RF filters and discusses the performance characteristics of the four basic types of filters, namely, low pass, high pass, band pass and band stop filters. Industrial applications of the most commonly used filters are indicated. The RF filter terms, Characteristic impedance, Transmission lines and VSWR are fundamental for all RF engineers and they are explained in detail in Chap. 2. The concept of impedance is derived from resistance using a simple example. The commonly used transmission line structures, coaxial, microstrip and stripline lines, are described. VSWR is explained graphically for open circuit, short circuit, matched and partially matched load conditions with numerical examples for better understanding.

RF filters are characterised by electrical specifications. The electrical specifications are explained in detail with random limiting values for low pass, high pass, band pass and band stop filters in Chap. 3. Typical performance graphs are provided for the four basic types of filters. Mechanical specifications and environmental requirements are industrial needs in designing RF filters. Mechanical specifications cover dimensions (outline), weight, input/output connectors and finish. Environmental specifications pertain to operating temperature, humidity and sealing requirements. The industrial needs are briefly presented.

Lumped/Semi-lumped, Microwave cavity and Microstrip/Stripline technologies for designing RF filters are presented in Chap. 4. Electrical circuit configurations are indicated for the lumped low pass, high pass, band pass and band stop filters explaining the functioning of the filters. The concept of semi-lumped filter is explained indicating the method of developing distributed capacitance. Structure, applications, advantages and the equivalent circuits of combline and iris-coupled microwave cavity filters are discussed. The construction and applications of basic microstrip/stripline circuits and computer aids for the design of the filters are briefly presented.

RF filter designs that just satisfy the specified requirements of customers are termed as functional designs. Functional design is definitely the core design task but it alone is not adequate. RF filters should also be designed for repeatability, reproducibility, produceability and reliability in addition to functional design requirements. Simple definitions for the four abilities, the benefits of value added design and the practical method of integrating the abilities with functional design are explained in Chap. 5.

The design of lumped and semi-lumped tubular low pass filters is illustrated with examples in Chap. 6. Guidance is provided for the selection of inductors and capacitors of lumped RF filters. The method of realising distributed capacitors and the interconnection of the reactive elements are explained for tubular filters. End coupling to input/output connectors and the tuning of filters are also explained.

The design of combline and iris-coupled cavity band pass filters is illustrated with examples in Chap. 7. Guidance is provided for the design of the piece parts of the cavity filters. The design graphs presented in the Microwave filter design book [1] need to be referred for designing cavity filters. Polynomials are fitted to the design graphs using computer-oriented numerical techniques [2], eliminating the need to refer the graphs. The methods of coupling end resonators to input/output connectors and the tuning of filters are explained.

Practical expertise is shared for the design and manufacturing of lumped low pass RF filter, semi-lumped tubular low pass filter, combline band pass filter and iris-coupled band pass filter in Chap. 8. Preventing the degradation of the reliability of piece parts of filters, tuning, 'cut and try' method, cross coupling for steep roll down filters and developing filter design software are some of the important topics covered in the chapter.

Additional information regarding the modes of transmission, RF coaxial connectors and statistical analysis of filter characteristics for six sigma design are provided in the Appendices.

References

1. Matthei GL, Leo Y, Jones EMT (1980) Microwave filters, impedance-matching networks, and coupling structures. Artech House, Norwood
2. Rajaraman V (1985) Computer oriented numerical techniques. Prentice Hall International, New Delhi

Acknowledgments

Throughout my career, I was assisted by a team of committed and talented engineers and technicians with a lot of enthusiasm and initiatives. I would like to thank the team first for making me a better professional with time.

I would like to thank my wife, Rameswari, for encouraging me to write the book after my retirement. I thank my son, Dr. Kumar Ph.D. for reviewing the manuscript with patience and providing me valuable suggestions. I also thank my son-in-law, Gopinathan, Semi-conductor designer, for technical support and my daughter, Sumithra, supporting me for figures.

Most importantly, I thank C. Krishnamurthy, Semi-conductor professional, General Manager (Retd.), Bharat Electronics, who ignited the opportunities for me to become a true professional in the field of Q&R. The expertise in Q&R enabled me to become a microwave professional later and to shine in my academic career.

Contents

Chapter 1
Introduction to RF Filters

Abstract The need for RF filters to attenuate interference band of frequencies, generated in signal processing circuits is explained with a practical example. The pass band of filters could be positioned in four ways and hence there are four types of filters. The basic characteristics of low pass, high pass, band pass and band reject filters are explained. Integrated filtering circuits and the industrial applications of filters are also explained with examples.

1.1 Need for Filters

Electronic equipments contain circuits for processing signal band of frequencies to satisfy the functional requirements of the equipments. For reasons such as non-linearity in the signal processing circuits and internal/external noise generation, frequency components outside the signal band are generated. The frequency components outside the signal band are the interference band of frequencies and they affect the satisfactory functioning of electronic equipments. They could cause distortion, jamming of received signals or poor reception in communication equipments. They could lead to erroneous diagnosis in medical equipments. A practical example is provided for the interference band of frequencies.

CDMA (Code-Division Multiple Access) services use 800 MHz band and the GSM (Global System for Mobile communication) services use 900 MHz band (GSM low band) and 1,800 MHz band (GSM high band). For simplicity, assume that CDMA system generates interference component in GSM low band at 920 MHz and the interference signal is shown in Fig. 1.1. The interference component causes the distortion or cross talk in GSM low band.

The interference component at 920 MHz in GSM band should be attenuated. A filter module is designed to attenuate the entire GSM low band and not just 920 MHz as the CDMA transmission system might generate other interference

D. Natarajan, *A Practical Design of Lumped, Semi-Lumped and Microwave Cavity Filters*, Lecture Notes in Electrical Engineering,
DOI: 10.1007/978-3-642-32861-9_1, © Springer-Verlag Berlin Heidelberg 2013

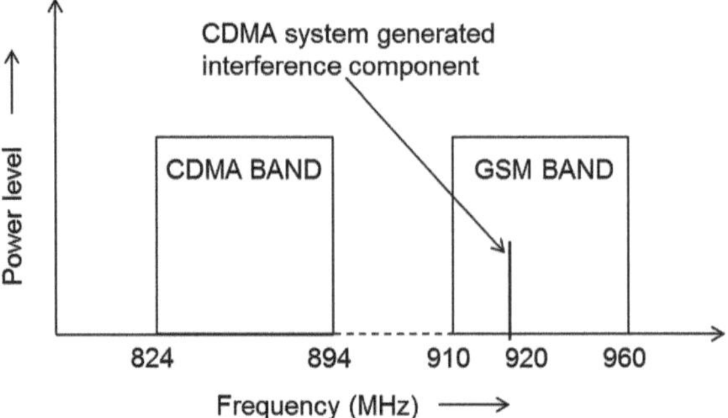

Fig. 1.1   Example of signal interference in bands

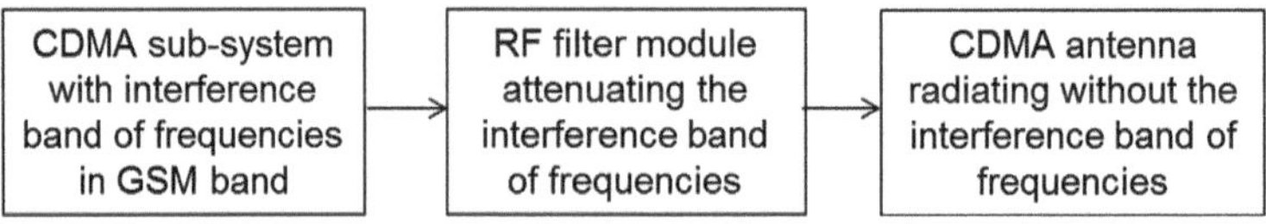

Fig. 1.2   Need for RF filter in electronic system

frequency components in GSM band. The filter module is inserted between CDMA sub-system and antenna as shown in Fig. 1.2 attenuating the interference band of frequencies.

1.2 Filtering Circuits

Simple requirements of suppressing interference signals are realised by integrating filter circuits in functional circuits. Two examples could be given. Fig. 1.3 shows a relay switching circuit. When the DC supply to the relay coil is switched off, reverse high voltage noise frequency components are generated as the inductance of the relay coil opposes 'change of current'. The noise frequency components are suppressed by connecting a diode in reverse direction across the relay coil.

The diode does not conduct or load the DC supply but act as a short circuit for the reverse high voltage noise frequency components thus filtering out the noise signals at the source itself.

Figure 1.4 shows the block diagram of a RF power amplifier with its inputs and output. Ceramic de-coupling capacitor is used across the DC supply input of the power amplifier to suppress or filter-out RF signal frequency components which otherwise would interfere with the functioning of DC supply.

The diode and the de-coupling capacitors are examples of simple RF filter circuits. However, attenuating the interference band of frequencies such as the requirements shown in Fig. 1.2 need dedicated RF filter modules.

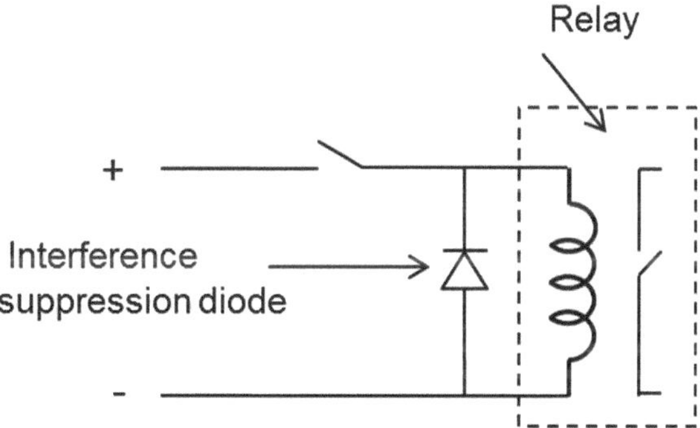

Fig. 1.3 Filtering circuit in relay switching

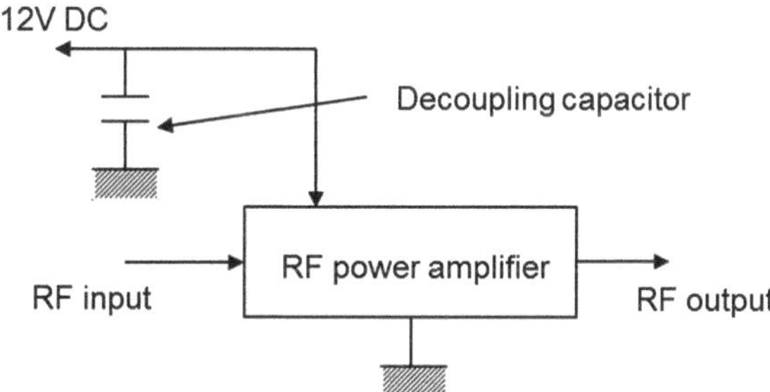

Fig. 1.4 Filtering circuit in RF amplifier

1.3   Types of Filters

Pass band of a filter is defined as the spectrum (range) of frequency components that pass through the filter with minimum attenuation or insertion loss. The positioning of pass band decides the type of filter. The pass band could be positioned in four ways and hence there are four types of RF filters. Low pass, High pass, Band pass and Band reject (stop) are the four types of filters, designed for various industrial applications. The basic characteristics of the filters are briefly explained.

1.3.1 Low Pass Filters

In low pass filter, the pass band is from DC to desired frequency. Pass band signals are allowed with minimum loss and the frequencies above the desired frequency are attenuated. Typical characteristic of a low pass filter is shown in Fig. 1.5 with f_u, the desired upper edge frequency of pass band. The frequency band above the frequency, f_u, is known as stop band. The starting frequency is shown as 10 MHz in Fig. 1.5 instead of DC and it is acceptable for low pass filters.

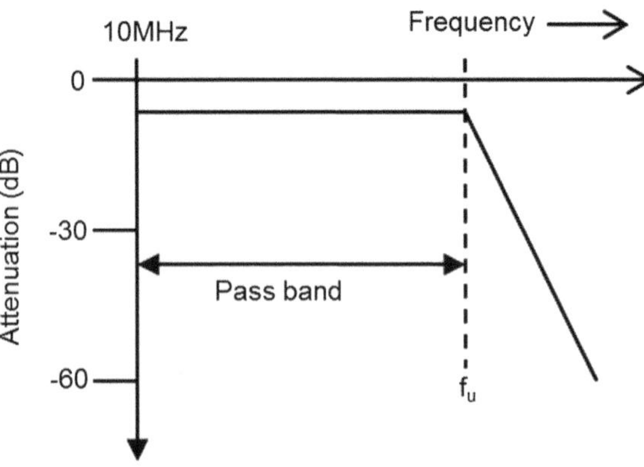

Fig. 1.5 Characteristic of low pass filter

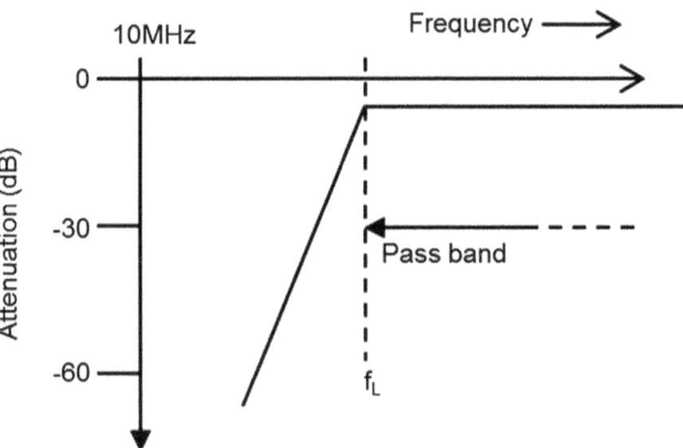

Fig. 1.6 Characteristic of high pass filter

1.3.2 High Pass Filters

In high pass filter, the pass band is from the desired frequency and above. Pass band signals are allowed with minimum loss and the frequencies below the desired frequency are attenuated. Typical characteristic of a high pass filter is shown in Fig. 1.6 with f_L, the desired lower edge frequency of pass band. The frequency band below the frequency, f_L, is the stop band.

1.3.3 Band Pass Filters

The pass band of band pass filters is defined by lower and upper limiting frequencies. These filters allow signals within the pass band with minimum loss but reject frequencies outside the pass band with maximum attenuation. Typical characteristic of a band pass filter is shown in Fig. 1.7 with lower and upper limiting frequencies, f_L and f_u, respectively. The frequency bands below and above

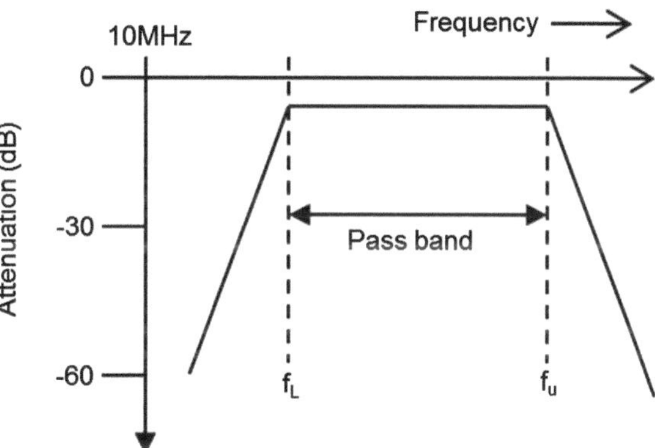

Fig. 1.7 Characteristic of band pass filter

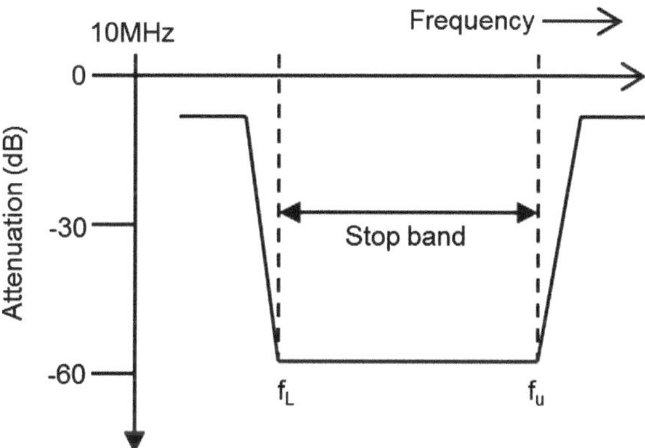

Fig. 1.8 Characteristic of band stop filter

the limiting frequencies are the stop bands. Band pass filters have one pass band and two stop bands.

1.3.4 Band Reject Filters

Band reject (stop) filters are used to attenuate a desired frequency band, defined by lower and upper limiting frequencies. Band stop filters reject frequencies within the desired band with maximum attenuation. The filters allow signals outside the desired band with minimum loss. Band stop filters have one reject band and two pass bands. One pass band is below the lower limit frequency and the other pass band is above the upper limit frequency. Band stop filters designed to attenuate a narrow frequency band with high attenuation are called Notch filters. Typical characteristic of a band stop filter is shown in Fig. 1.8 with lower and upper limit frequencies, f_L and f_u, respectively.

1.4 Application of RF Filters

Most of the microwave applications need band pass filters [1]. For example, the filter shown in Fig. 1.2 requires a band pass filter for the 800 MHz band with sharp rolling down of pass band from upper limit frequency. Band pass filters are used in base station transmitters. Ultra-wideband (UWB) band pass filters with bandwidth of 7.5 GHz for the unlicensed spectrum from 3.1 to 10.6 GHz for many applications such as short-range high-data rate communication systems and wireless personal area networks [2, 3]. Low pass filters are used in communication systems to suppress spurious modes in oscillators and leakage in mixers [4]. Tunable notch filters and tunable band pass filters are used in military communication systems. Filters are also used for selecting a band of frequencies.

References

1. Hunter I, Ranson R, Guyette A, Abunjaileh A (2007) Microwave filter design from a systems perspective. IEEE Microw Mag 8:71–77
2. Hao ZC, Hong JS (2010) Ultrawideband filter technologies. IEEE Microw Mag 11:56–68
3. Stark A, Friesicke C, Muller J, Jacob AF (2010) A packaged ultrawideband filter with high stopband rejection. IEEE Microw Mag 11:110–117
4. Hettak K, Stubbs MG, Elgaid K, Thayne IG (2005) A compact, high performance, semi-umped, low-pass filter fabricated with a standard air bridge process, MW conference, 2005 European, pp 4–7

Chapter 2
RF Filter Terms

Abstract Impedance, Characteristic impedance, RF transmission line and VSWR are fundamental terms not only for the design of RF filters but also for all RF components and circuits. The terms are explained in detail with simple examples for physical understanding. Impedance is explained for a discrete reactive element circuit and for a single load resistor introducing the concept of distributed reactive elements. The construction and characteristics of coaxial, microstrip and stripline transmission lines are explained. VSWR is explained graphically with numerical examples for transmission line with open-circuited load, short-circuited load, matched load, infinitely long coaxial cable and partially matched load after explaining its practical significance.

2.1 Introduction

RF filters are designed as per customer requirements. The requirements vary among the four basic types (low pass, high pass, band pass and band stop) of filters. However, all the four types of filters have a few requirements in common. RF filters are designed for the standardised input/output characteristic impedance, 50 Ω. RF transmission lines are used in the design of microwave filters. Input and output VSWR of filters are measured to verify compliance to the specified characteristic impedance and to ensure minimum insertion loss. Impedance, Characteristic impedance, RF transmission lines and VSWR are fundamental terms not only for the design of RF filters but also for all RF components and circuits. Hence, the terms are explained in detail with examples. The other filter terms such as Insertion loss, Band width and Rejection represent the characteristics of filters and they are explained in Chap. 3.

D. Natarajan, *A Practical Design of Lumped, Semi-Lumped and Microwave Cavity Filters*, Lecture Notes in Electrical Engineering,
DOI: 10.1007/978-3-642-32861-9_2, © Springer-Verlag Berlin Heidelberg 2013

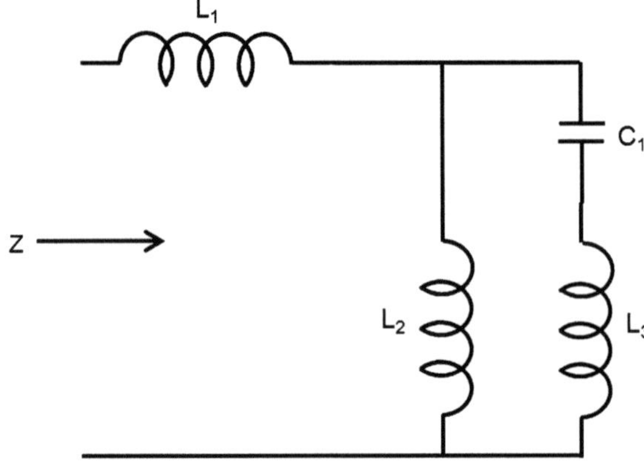

Fig. 2.1 A discrete reactance circuit

2.2 Impedance

2.2.1 Definition

Resistance (R) is defined as the measure of the opposition to direct current (DC) or alternating current (AC) assuming that the resistance is pure in the sense that it does not have inductive or capacitive reactance. It is expressed in ohms.

Impedance (Z) is analogous to resistance and is defined as the measure of the opposition to the flow of alternating current. Resistance (R) and reactance (X) are part of impedance. Reactance may be due to inductance or capacitance or both. It is expressed in ohms.

$$Z = \sqrt{(R^2 + X^2)}$$

If $X = 0$, the impedance is said to be resistive. At lower frequencies, impedance is relevant for a discrete circuit having resistors, inductors and capacitors connected in series/parallel configuration. A 4-element discrete circuit with capacitors and inductors is shown in Fig. 2.1. The impedance of the circuit is calculated using appropriate expressions for the series/parallel configuration.

2.2.2 Load Resistor at Radio Frequency

At VHF and microwave frequencies, impedance becomes relevant even for a single element circuit having one resistor. Figure 2.2 shows a single element load resistance (R_L) circuit with a VHF signal source having source impedance, Z_S. Though inductors and capacitors are not physically connected in the circuit, the load resistance is load impedance to the VHF source due to the effect of distributed inductance and capacitance present in the load resistor, R_L. The distributed elements could be understood by examining the construction of the load resistor.

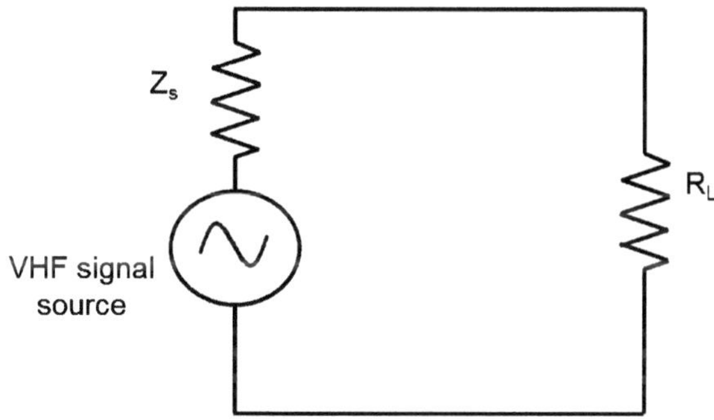

Fig. 2.2 VHF circuit with load resistance

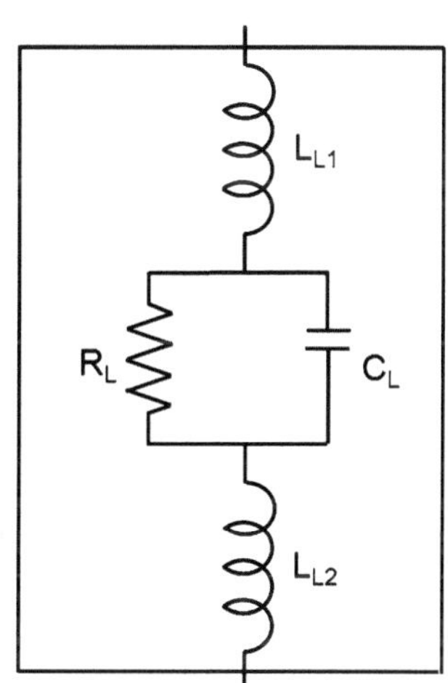

Fig. 2.3 RF equivalent circuit of the load, R_L

For better understanding, assume that the load resistance, R_L, is a resistor with leads. Leaded resistor has a ceramic rod over which the resistive element is vacuum deposited. The deposited resistive element is spirally cut to adjust its resistance value within tolerance. Metallic end caps are attached to the ceramic rod and they make electrical contact with the resistive element. The leads (terminations) are then attached to the end caps. Finally, a suitable insulating epoxy is applied over the resistive element for environmental protection.

Examining the construction of the load resistor, the terminations function as inductors, L_{L1} and L_{L2}. The protective epoxy coating acts as a dielectric and hence a capacitor, C_L, is present between the end caps. L_{L1} and L_{L2} are distributed inductances and C_L is distributed capacitance as they are spread (distributed) over the leaded load resistor. Other distributed reactive elements in the load resistor are negligible. The RF equivalent circuit of the load resistor with its resistive and distributed reactive elements is shown in Fig. 2.3. Hence, the load resistance, R_L, with its distributed reactive elements becomes load impedance, Z_L at VHF and microwave frequencies.

2.2.3 Standard Terminations

For the performance verification of RF components such as filters, the input of the component is connected to RF source and the output is terminated with resistive load impedance. Load impedances are standardised to 75 and 50 Ω. Standard termination is a commercial terminology for the standardised loads. The terminations are specially designed coaxial film resistors minimising distributed capacitance and inductance. They are available in BNC, TNC, N, SMA and other connector series. The terminations are used for the initial set up calibration of RF test equipments. The frequency of application is generally limited to 1 GHz for 75 Ω terminations whereas 50 Ω terminations are used up to microwave frequencies.

The concept of minimising distributed inductance and capacitance is used to design resistive termination at microwave frequencies by cancelling inductive reactance with capacitive reactance at that frequency using reactive matching networks [1].

2.3 RF Transmission Line

2.3.1 Definition

A transmission line is a pair of electrical conductors that transfer power from a source to a load. One of the two conductors functions as forward path from the source to the load and the other functions as return path from the load to the source thus closing the circuit for current flow. A transmission line with open type of conductors is shown in Fig. 2.4.

Transmission lines with open type conductors function satisfactorily for transferring power from source to load at DC and at power frequency (50/60 Hz)

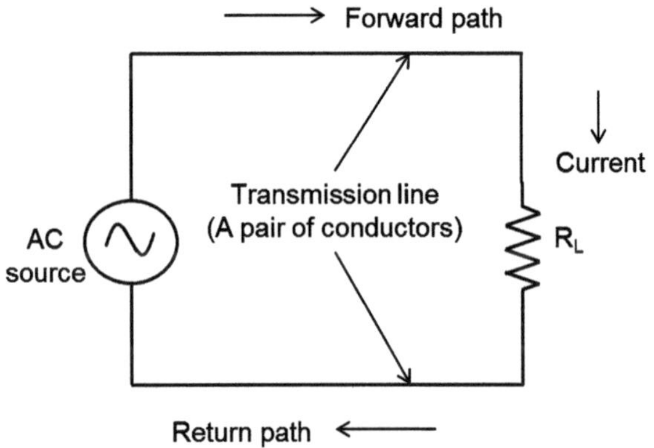

Fig. 2.4 Open type transmission lines

or up to 100 kHz with acceptable degradation. However, the open transmission lines are not acceptable at radio frequency as it will result in the loss of RF signals when transferring power from source to load. RF signals are electromagnetic waves and they propagate through a dielectric medium including free space and vacuum. A specially designed structure having a dielectric medium contained by electrically conducting walls is required for the propagation of electromagnetic waves to minimise the loss of RF signals. The specially designed structure is RF transmission line. Many designs of transmission lines are available considering frequency, loss and other interconnection requirements of applications.

2.3.2 Types of Transmission Lines

The RF transmission line structures that find wide applications are:

1. Coaxial transmission lines
2. Microstrip transmission lines
3. Stripline transmission lines

Feeder cables to antennas and interconnection cables between RF sub-systems are some of the applications of coaxial cables. Microstrip and stripline lines find applications in the design of filters and other RF components. The construction and characteristics of coaxial and microstrip/stripline transmission lines are explained. Waveguides are special form of transmission lines and they are used at microwave frequencies.

2.4 Coaxial Transmission Lines

2.4.1 Construction and Characteristics

Coaxial cables are the most commonly used form of coaxial transmission lines. The cables are flexible and they are available off-the-shelf with varied constructions. In a coaxial cable, the inner and outer conductors share the same axis i.e. they are designed to be coaxial to each other. The cross section of a coaxial cable in general form is shown in Fig. 2.5.

A dielectric material is placed between the inner and outer conductors to maintain the coaxial structure throughout its length. The dielectric material is polyethylene or PTFE in coaxial cables. The electromagnetic waves of RF signal are contained between inner and outer conductors and they travel in the dielectric medium. The constructional features of coaxial cables decide the attenuation (loss), maximum frequency of application and operating ambient temperature.

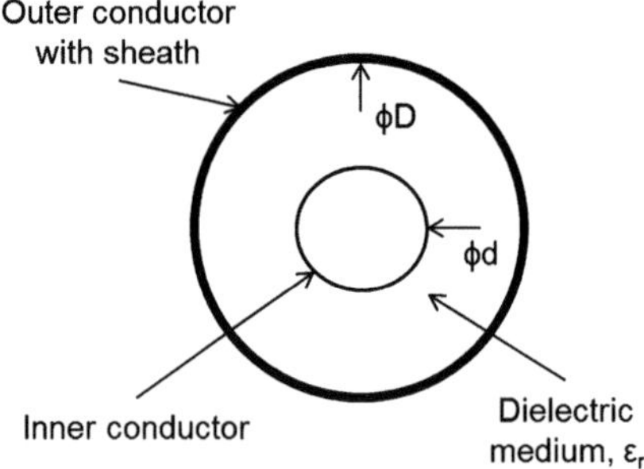

Fig. 2.5 Cross section of a coaxial cable

The dielectric material of coaxial cables contributes significantly for the loss of the cables. Air bubbles are injected during the manufacture of polyethylene or PFFE dielectric material and the air-foamed dielectric cables have lower loss. Selecting coaxial cables with higher diameter or with silver plated inner/outer conductors reduces the loss of the cables further. Coaxial cable is available with outer conductor braided with many strands of copper wires or with seamless copper tube. Coaxial cables with seamless copper tube outer conductor and PTFE dielectric are semi-rigid cables and they are suitable for applications up to 20 GHz. The maximum operating ambient temperature is 85 °C for cables with polyethylene dielectric and is 180 °C for PTFE dielectric cables.

TEM (Transverse Electro-Magnetic) mode is the dominant mode of propagation in RF coaxial cables. The propagation modes of transmission is briefly explained in Appendix-1. TEM mode in coaxial cables changes to higher order modes at a frequency called cut-off frequency, resulting in very high attenuation and hence the operating frequency of coaxial cables should be lower than the cut-off frequency. The expression published by leading international cable manufacturers is useful to estimate the cut-off frequency, f_c, of a coaxial cable.

$$f_c(GHz) \approx 191/[(D + d)\,(\sqrt{\varepsilon_r})]$$

D   Inner diameter of outer conductor in mm
d   Outer diameter of inner conductor in mm
ε_r is the relative dielectric constant of the medium between the conductors

2.4.2 Standard Coaxial Air Lines

RF coaxial transmission lines with air as dielectric are Standard air lines. Standard air line has rigid silver plated inner and outer conductors which are connected to precision 7 or 3.5 mm coaxial connectors on both sides. The centre conductor of standard air line is mechanically supported by the centre conductors of the end coaxial connectors. Standard air lines are characterised by the lowest loss per meter and VSWR. They are used as standards for impedance measurements.

2.5 Microstrip/Stripline Transmission Lines

Microstrip or stripline transmission line patterns are designed and printed on PTFE copper clad laminate. The laminate is a sheet of PTFE insulating material (substrate) bonded with copper sheet on both sides. Transmission line patterns printed on the laminate by special chemical etching processes. The unwanted copper is etched out on both sides of substrate by the chemical processes. Copper conductor (transmission line) width, thickness of substrate and the dielectric constant of the substrate decide the 'characteristic impedance' of microstrip or stripline transmission lines. Characteristic impedance and the applications of basic microstrip and stripline configurations are explained in Chap. 4. Advanced filler materials are added to the PTFE substrate for improving the stability and Q-factor of PTFE copper clad laminate. For HF and VHF applications, glass epoxy copper clad sheets is also used for designing RF components in microstrip and stripline configurations.

2.6 Characteristic Impedance

Characteristic impedance is explained for coaxial transmission line. The cross section of a coaxial cable shown in Fig. 2.5 is referred for explaining the characteristic impedance of a coaxial transmission line. Assume that the length of coaxial cable is infinite or long enough to act as load impedance. The inner and outer conductors have series distributed inductance. Distributed capacitance is present across the conductors throughout the length of the coaxial cable. The cable has also series DC resistance and shunt conductance but their contributions for the input impedance of the cable are considered negligible compared to the reactance at high frequencies. The input impedance of the cable is given by the expression, $\sqrt{(L/C)}$, where L is the distributed inductance and C is the distributed capacitance per unit length of the coaxial cable [2]. The values of L and C are related to the parameters, D, d, and ε_r, shown in Fig. 2.5. In other words, D, d and ε_r distinctly defines or characterises the input impedance of a coaxial cable. Hence, the characteristic impedance of a coaxial transmission line is defined as the input impedance of the line and is expressed in terms of D, d and ε_r [3]. Its unit is ohms and the symbol is Zo.

$$\text{Characteristic impedance}, Z_0 = \frac{60\ln(D/d)}{\sqrt{\varepsilon_r}} \text{ ohms}$$

Characteristic impedance of a coaxial cable is measured using Vector Network Analyser with Time Domain Reflectometer (TDR) software. TDR equipment with Standard air line is also used for measuring the impedance. As the characteristic impedance of a coaxial cable is related to its dimensions, the bending diameter of the cable should exceed 10 times the cable diameter during handling or applications.

The loss (attenuation) of RF transmission lines is minimum at Zo = 77 Ω and the power handling capacity of the lines is maximum at Zo = 30 Ω. Considering low loss requirements, characteristic impedance of 75 Ω is standardised for video applications. Characteristic impedance of 50 Ω is standardised for all other applications balancing loss and power handling requirements [4].

2.7 VSWR

2.7.1 Maximum Power Transfer

The acronym, VSWR, stands for Voltage Standing Wave Ratio. Figure 2.6 shows a series circuit having RF signal generator (source), a short length of RF coaxial transmission line and load impedance. The internal (output) impedance (Z_s) of the signal source is also shown. The transmission line is shown thick for simplicity.

Maximum power from the source is transferred to the load impedance when the output impedance of the RF signal source, characteristic impedance of the transmission line and the load impedance are matched. Assume that the output impedance (Z_s) of the RF signal source and the impedance of the transmission line (coaxial cable) are 50 Ω. The power from the source is defined as incident power. If the load impedance in the RF circuit is also 50 Ω i.e. matched to the impedance of transmission line and the RF source, maximum power transfer occurs from source to load. The incident power travels through the transmission line towards the load and it is fully absorbed by the load under matched conditions. If the source impedance is not resistive, then the load impedance must be complex conjugate of the source impedance for maximum power transfer. For example, if the source impedance is (30 + j2) Ω, then the characteristic impedance of the transmission line must be (30 + j2) Ω and the load impedance must be (30 − j2) Ω for maximum power transfer.

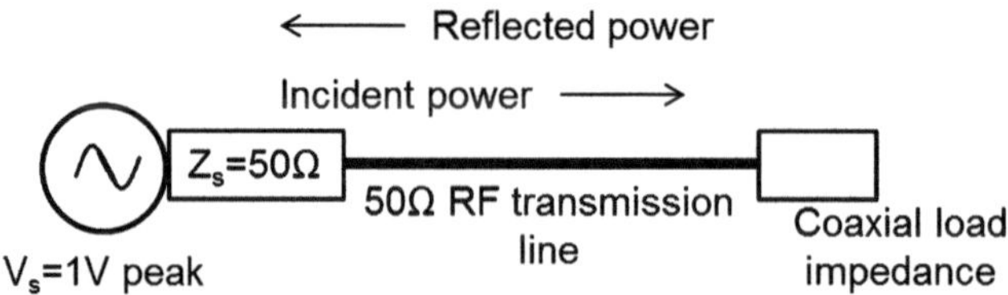

Fig. 2.6 RF transmission line circuit

If the load impedance is not matched to the impedance of RF circuit, some percentage of the incident power is returned i.e. reflected by the load towards the source through the transmission line. The quantum of the reflected power depends on the level of mismatch. If the load is open-circuit (load not connected to the transmission line) or the load is short circuit, the incident power is totally reflected at the end of the transmission line. The end of the transmission line is termed as the plane of the load. If the load impedance is partially matched (Ex: 40 or 60 Ω), then a fraction of the incident power is reflected back towards the source and the remaining power is absorbed by the load. The directions of incident power and the reflected power are shown in Fig. 2.6. Practical significance of VSWR is explained with an example.

2.7.2 Practical Significance of VSWR

Assume that a RF amplifier feeds an antenna through a RF coaxial feeder cable. The power amplifier is the RF source and the feeder cable is the transmission line. The antenna serves as the load impedance. Let the output impedance of the power amplifier and the characteristic impedance of the feeder cable is 50 Ω. Let the impedance of the antenna is 35 Ω, not matched to that of the source and the feeder cable. Due to impedance mismatch, a portion of the incident power is reflected by the antenna towards the amplifier and the remaining power is radiated. The reflected power is dissipated in the power amplifier and the dissipation could result in the premature (degradation) failure of the RF power transistor of the power amplifier. Customers experience poor reception due to the radiation of less power by the antenna. Hence, impedance matching is important for the satisfactory performance of RF systems. VSWR is a metric that measures how well the impedances of various components are matched in RF circuits and systems to realise maximum power transfer. VSWR has no units as it is a ratio of voltages.

VSWR is explained for four cases of load impedance i.e. open circuit, short circuit, matched (50 Ω) and partially matched (Ex.: 40 or 60 Ω), assuming the output impedance of the RF signal source and the impedance of the transmission line (coaxial cable) are 50 Ω. It is also assumed that the transmission line is *lossless* and the source delivers a sinusoidal voltage, V_s, of 1 V peak at microwave frequency, f (wave length, λ).

2.8 VSWR: Open Circuited Load

The transmission line circuit in Fig. 2.6 is modified and redrawn without load in Fig. 2.7, showing only one wave length (λ) of the RF transmission line and markers for every $\lambda/8$ from the load end. Assuming that the incident voltage wave

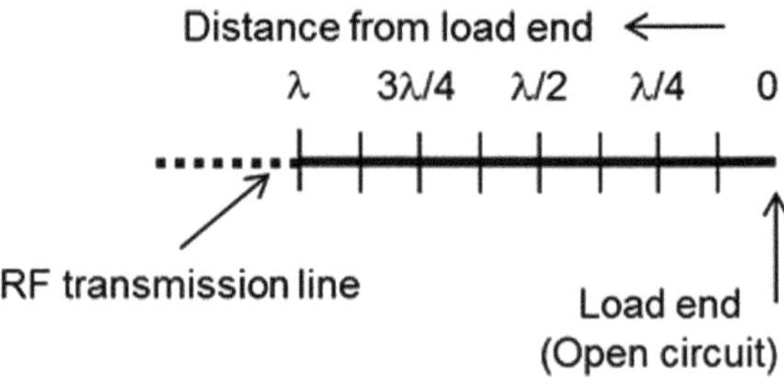

Fig. 2.7 RF transmission line—open circuit

originates at a distance, λ, from the load end, the analysis of incident and reflected waves is presented for one wave length.

2.8.1 Relationship Between Incident and Reflected Voltages

The incident voltage wave sees an open circuit at the plane of the load and hence it is totally reflected by the load. The current at the end of the transmission line is zero. The frequency of the reflected voltage wave is same as that of the incident voltage wave. The magnitude and sign of the reflected wave are also same as those of the incident wave at the plane of the load so that the current is zero. For example, if the instantaneous incident voltage is -0.7 V at the plane of the load, the reflected voltage is also -0.7 V at the plane of the load. The reflected wave is in-phase with the incident wave.

The instantaneous voltages of incident and reflected waves are analysed using the in-phase relationship between the waves. For example, if the instantaneous incident voltage wave rises positively from 0 V at the point λ, the instantaneous reflected voltage wave also rises positively from 0 V at the load end. If the incident voltage wave falls negatively from -0.7 V at the point λ, the instantaneous reflected voltage wave also falls negatively from -0.7 V at the load end. The instantaneous voltages of incident and reflected waves are presented at eight different times.

2.8.2 Instantaneous Voltages of Incident, Reflected and Resultant Waves

To begin with, i.e. at time, t = 0, the instantaneous voltages of incident and reflected voltage waves having a peak voltage of 1 V are shown in Fig. 2.7a in dashed lines. The instantaneous voltages of incident and reflected waves are vectorially added and the resultant voltage waveform is also shown in Fig. 2.7a in continuous line. The voltage of the resultant wave is zero.

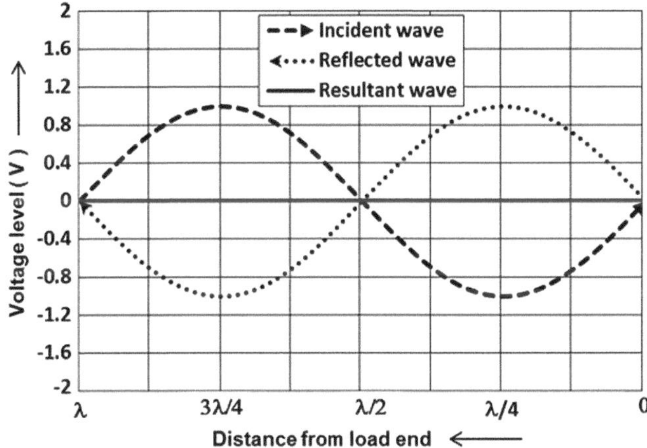

Fig. 2.7a   Instantaneous voltages at t = 0 for open circuited line

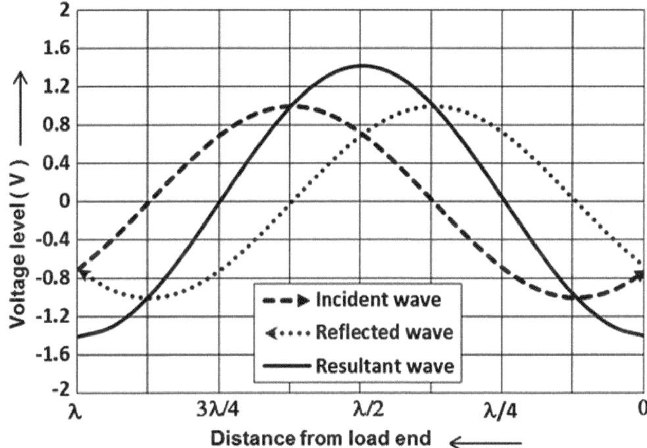

Fig. 2.7b   Instantaneous voltages at t = λ/8 for open circuited line

At t = λ/8, the incident voltage wave advances by λ/8 towards the load and the reflected voltage wave also advances by λ/8 towards the source. The wave forms of incident voltage, reflected voltage and the resultant voltage at t = λ/8 are shown in Fig. 2.7b with the levels of instantaneous voltages. The peak voltage of the resultant voltage wave varies from -1.4 to $+1.4$ V.

Similarly, the instantaneous voltage levels of incident, reflected and the resultant waveforms are obtained at t = λ/4, t = 3λ/8, t = λ/2, t = 5λ/8, t = 3λ/4, t = 7λ/8 and t = λ and they are shown in Fig. 2.7c–h respectively. It could be observed that the peak value of the resultant voltage waveform always occurs at λ/2 or its multiple from the open-circuited load end. Hence, the resultant wave is

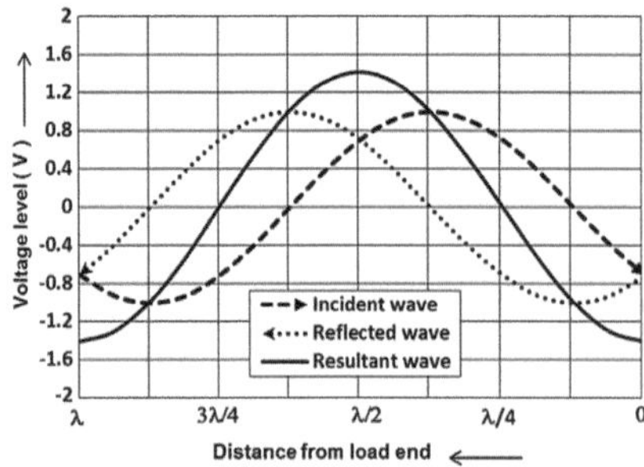

Fig. 2.7c Instantaneous voltages at t = $\lambda/4$ for open circuited line

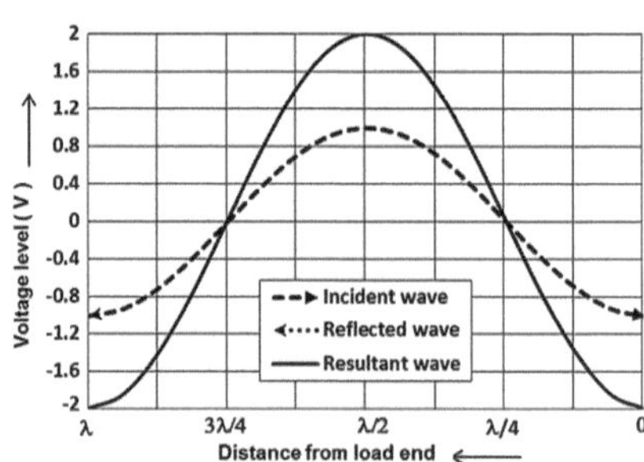

Fig. 2.7d Instantaneous voltages at t = $3\lambda/8$ for open circuited line

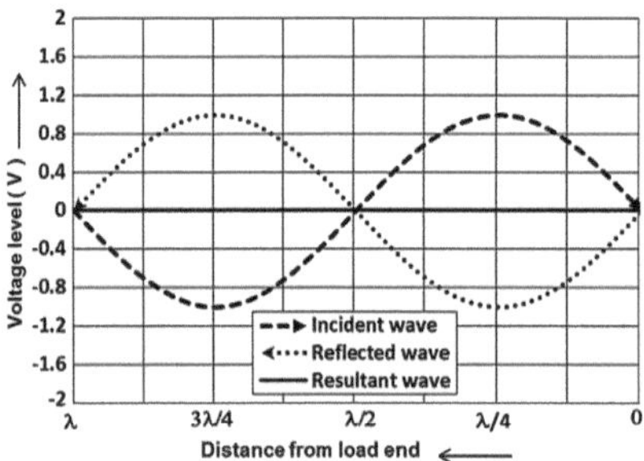

Fig. 2.7e Instantaneous voltages at t = $\lambda/2$ for open circuited line

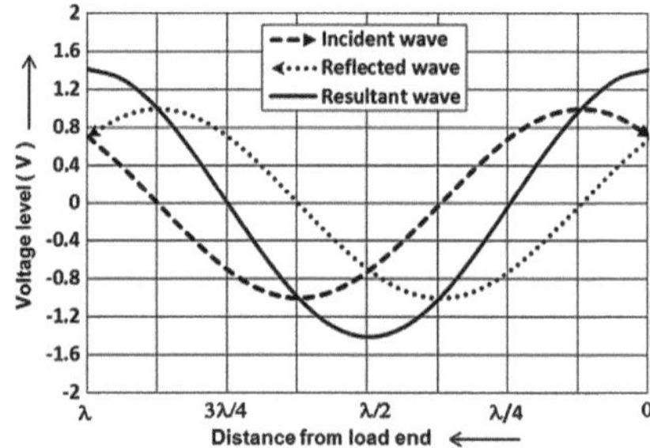

Fig. 2.7f Instantaneous voltages at t = $5\lambda/8$ for open circuited line

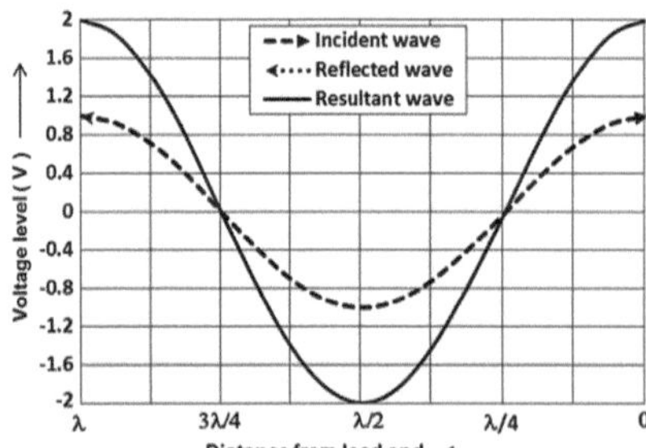

Fig. 2.7g Instantaneous voltages at t = $3\lambda/4$ for open circuited line

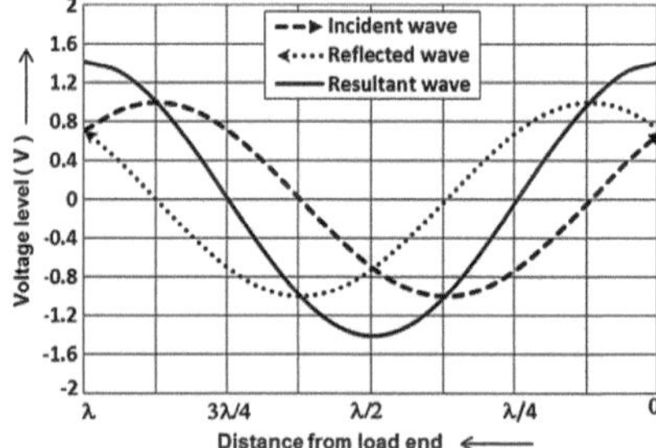

Fig. 2.7h Instantaneous voltages at t = $7\lambda/8$ for open circuited line

considered stationary i.e. standing. The peak voltage of the standing wave varies from -2 to $+2$ V.

2.8.3 Standing Voltage Wave in Slotted Line

The obsolete Hewlett-Packard slotted line is quite useful in demonstrating the standing wave along the transmission line [5]. The slotted line is a specially designed precision rigid coaxial transmission line with air as dielectric. It has a narrow longitudinal slot along the outer conductor of the line. A capacitive probe is inserted into the slot for voltage sampling. There is a mechanical carriage arrangement to move the probe along the slot without contacting the walls of the slot. Moving the probe along the slot is equivalent to moving it along the transmission line.

Assume the slotted line set-up is connected to the open-circuited transmission line, which has standing voltage waveform varying from -2 to $+2$ V. The sampled voltage signal from the probe is fed to an effective (rms) voltage reading voltmeter. If the probe carriage is moved from the open circuited load end of the slotted line towards the RF signal source, the contour of the effective voltage of the standing wave would be as shown in Fig. 2.8 with its voltage level varying from 0V to 1.4 V for the lossless transmission line.

The minimum voltage, V_{min}, is zero and the maximum voltage, V_{max}, is 1.4 V rms for the standing wave voltage. Minimum voltage points are termed as nodes and maximum voltage points are termed as antinodes [2]. The ratio of maximum and minimum voltages of the standing wave is defined as VSWR. VSWR of the transmission line at the point of discontinuity i.e. open circuit is:

$$VSWR = V_{max}/V_{min} = 1.4\,V/0\,V = \infty$$

In practice, VSWR has a finite value V_{max} is less than 1.4 V and V_{min} is more than 0 V as transmission line has a finite loss. In the hp slotted line set-up, the sampled voltage output of the probe is connected to SWR meter, which, in combination with a square-law detector, indicates the measured VSWR directly.

Fig. 2.8 Standing wave voltage pattern for open circuited line

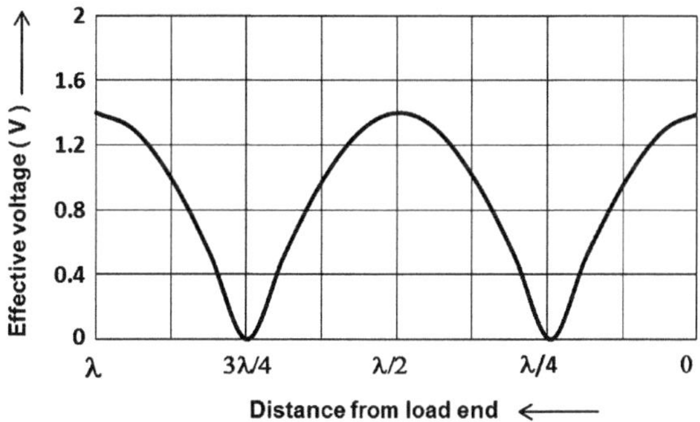

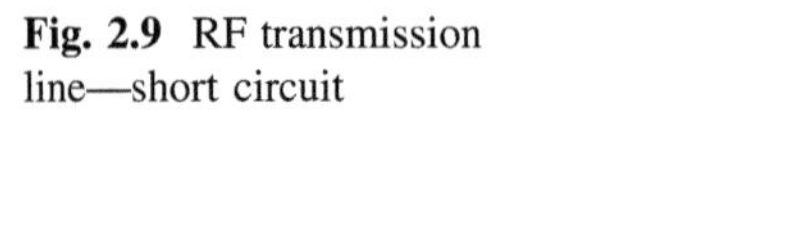

Fig. 2.9 RF transmission
line—short circuit

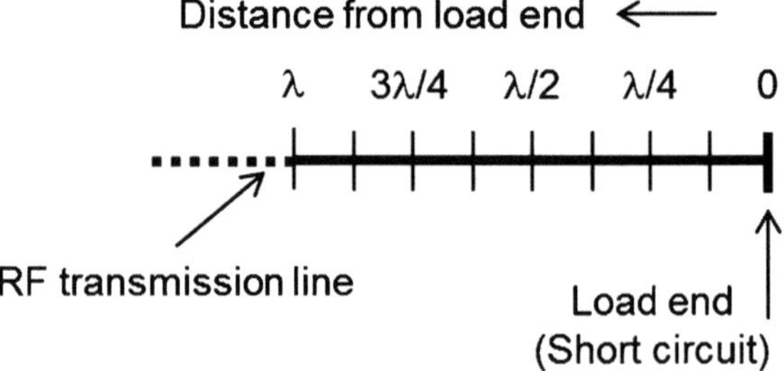

2.9 VSWR: Short Circuited Load

The RF transmission line with its load end short circuited is shown in Fig. 2.9 for one wave length (λ) from the load end with markers for every $\lambda/8$. The short circuited end is shown by a thick line perpendicular to the transmission line. The concept of standing voltage wave generation is basically same as that of open circuited transmission line.

2.9.1 Relationship Between Incident and Reflected Voltages

The incident voltage wave sees a short circuit at the plane of the load, and hence it is totally reflected by the load. The voltage at the end of the transmission line is zero. The frequency of reflected voltage wave is same as that of incident voltage wave. The magnitude of the reflected voltage wave is also same as that of incident voltage wave but the voltages oppose each other at the load end so that the resultant voltage is zero. For example, if the instantaneous incident voltage is −0.7 V, the reflected voltage is +0.7 V at the plane of the load. The reflected voltage wave is out of phase with the incident wave.

The instantaneous voltages of incident and reflected waves are analysed using the out-of-phase relationship between the waves. For example, if the instantaneous incident voltage wave rises positively from 0 V at the point λ, the instantaneous reflected voltage wave falls negatively from 0 V at the load end. If the incident voltage wave falls negatively from −0.7 V at the point λ, the instantaneous reflected voltage wave rises positively from +0.7 V at the load end. The instantaneous voltages of incident and reflected waves are presented at eight different times.

2.9.2 Instantaneous Voltages of Incident, Reflected and Resultant Waves

The instantaneous voltage levels of incident, reflected and the resultant waves for t = 0 to t = λ with intervals of $\lambda/8$ are shown in Fig. 2.10a–h. It could be observed

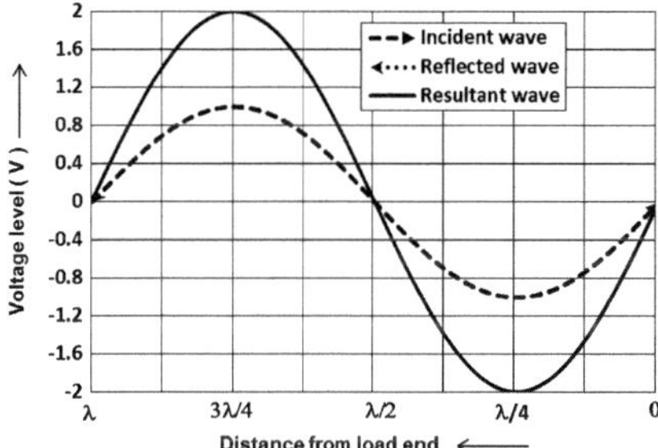

Fig. 2.10a   t = 0 for short circuited line

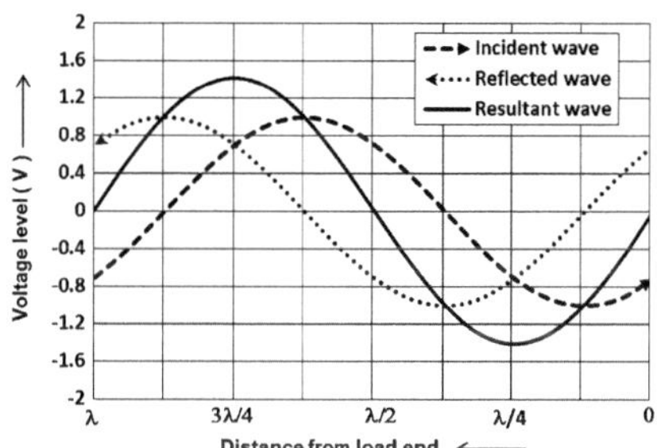

Fig. 2.10b   t = $\lambda/8$ for short circuited line

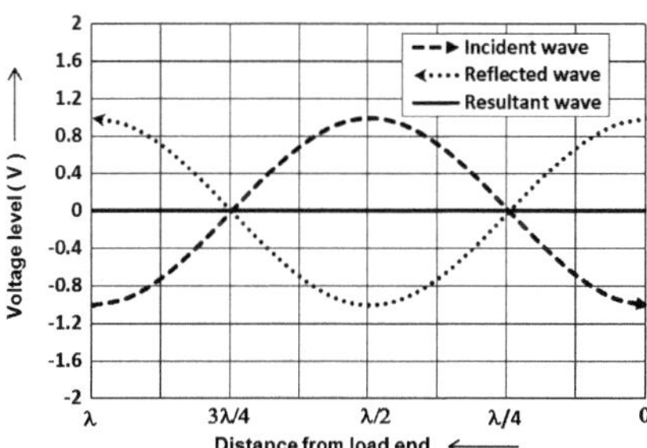

Fig. 2.10c   t = $\lambda/4$ for short circuited line

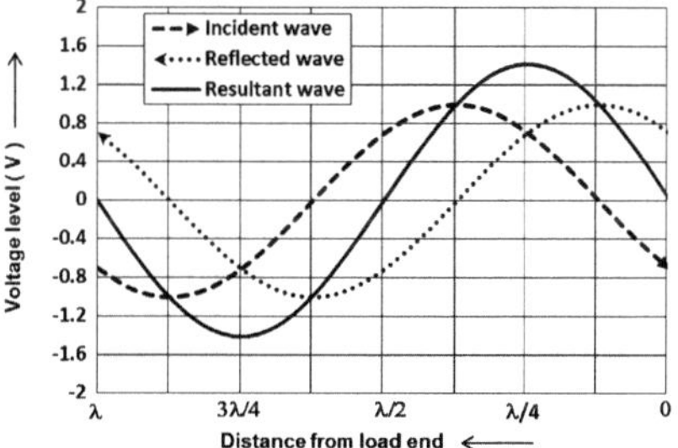

Fig. 2.10d   t = $3\lambda/8$ for short circuited line

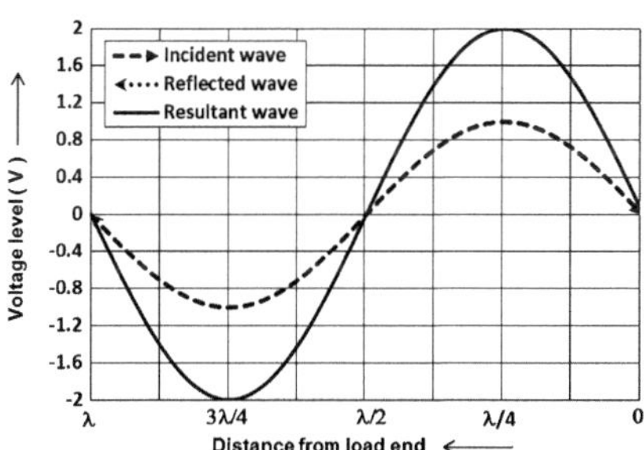

Fig. 2.10e   t = $\lambda/2$ for short circuited line

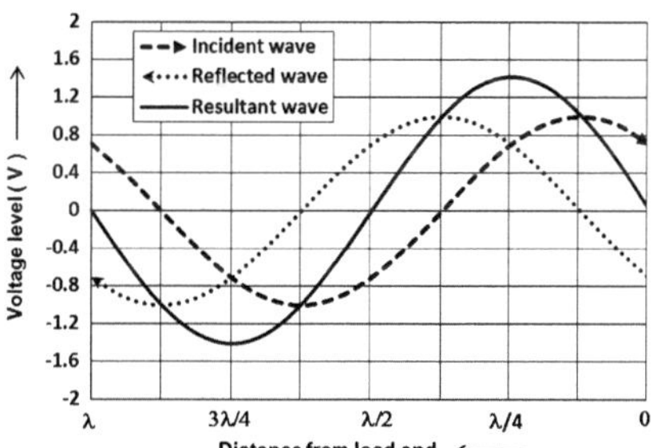

Fig. 2.10f   t = $5\lambda/8$ for short circuited line

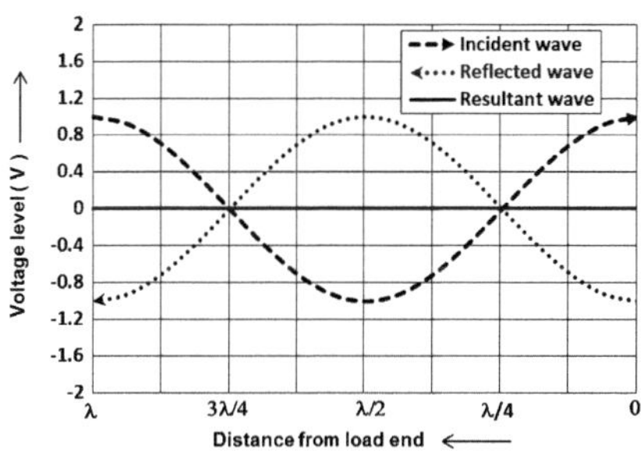

Fig. 2.10g   t = $3\lambda/4$ for short circuited line

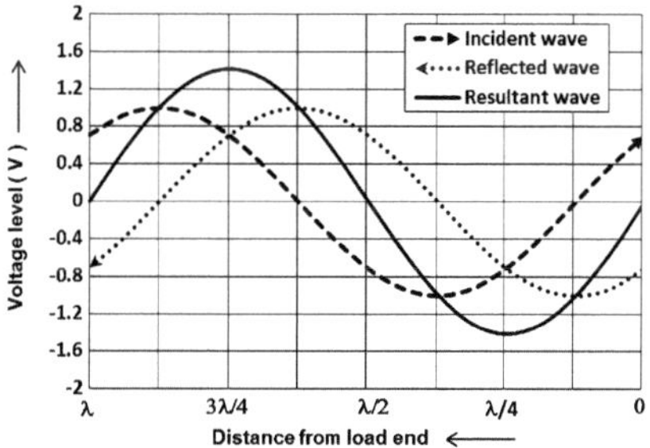

Fig. 2.10h   t = $7\lambda/8$ for short circuited line

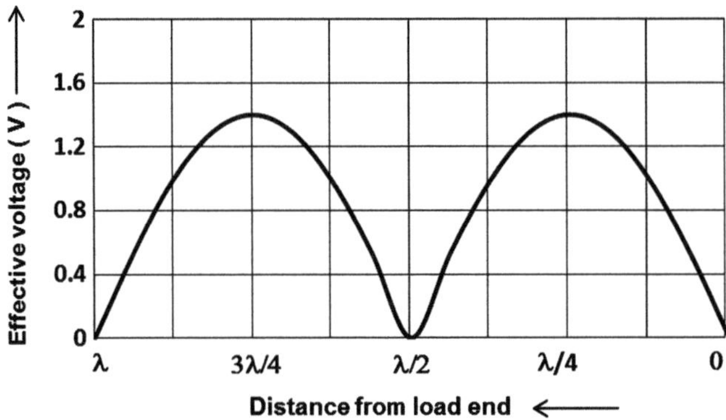

Fig. 2.11 Standing wave voltage pattern for short circuited line

that the peak value of the resultant standing voltage waveform always occurs at $\lambda/4$ or its multiple from the short-circuited load end. The peak voltage of the standing wave varies from -2 to $+2$ V.

In a slotted line set up, the contour of the effective voltage of the standing wave would be as shown in Fig. 2.11 with its voltage level varying from 0 V to 1.4 V for the lossless transmission line. For the RF transmission line with short circuit,

$$\text{VSWR} = V_{max}/V_{min} = 1.4\,V/0\,V = \infty$$

In practice, VSWR has a finite value V_{max} is less than 1.4 V and V_{min} is more than 0 V as transmission line has a finite loss.

2.10 VSWR: Matched 50 Ω Load

The RF transmission line, matched with 50 Ω load, is shown in Fig. 2.12 for one wave length (λ) from the load end. Markers are shown for every $\lambda/8$.

At the plane of the load, the incident voltage wave sees matched load and hence the incident wave is fully absorbed by the load. There is no reflected voltage wave. If an effective voltage reading voltmeter is moved along the transmission line from the load end, it reads a constant voltage, equal to the voltage across the load. For the source voltage of 1 V peak, the voltage across the load is 0.5 V peak and hence the rms voltmeter reads a constant voltage of $0.5/\sqrt{2}$ i.e. 0.35 V along the line as shown in Fig. 2.13.

$$V_{max} = V_{min} = 0.35\,V$$

$$\text{VSWR} = V_{max}/V_{min} = 1 \text{ (Ideal value)}$$

In practice, VSWR is always greater than the ideal value as transmission line has a finite loss. Hence, the rms voltmeter of slotted line shows the presence of a

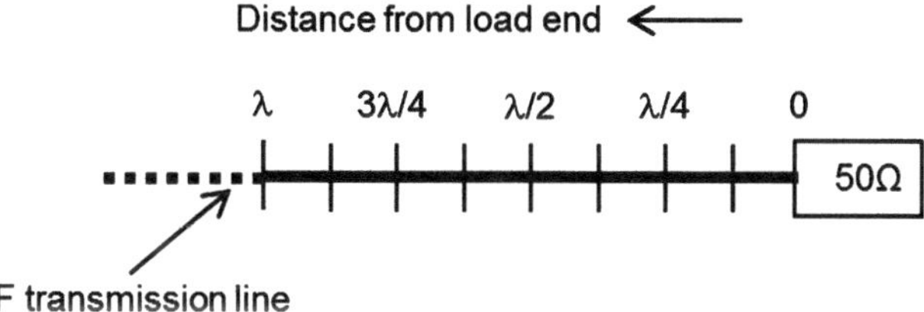

Fig. 2.12 RF transmission line—matched load

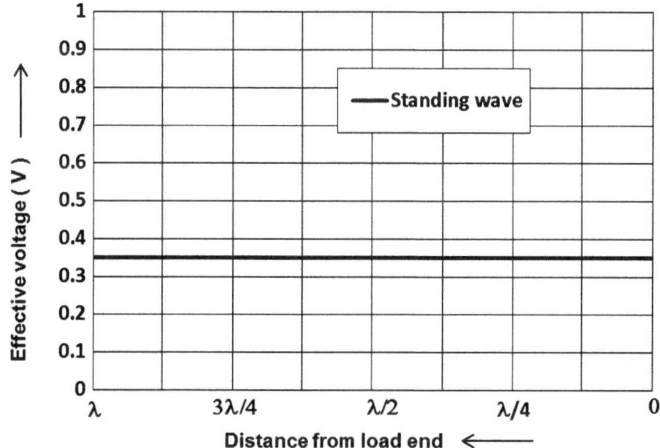

Fig. 2.13 Standing wave for transmission line with 50 Ω load

small ripple voltage instead of constant voltage. The precision hp slotted line explained earlier has a VSWR of less than 1.04.

2.11 VSWR: Infinitely Long 50 Ω Cable as Load

Figure 2.14 shows an infinitely long 50 Ω coaxial cable, connected to a transmission line instead of 50 Ω load impedance. Assume that the cable has a uniform characteristic impedance of 50 Ω and finite loss per meter.

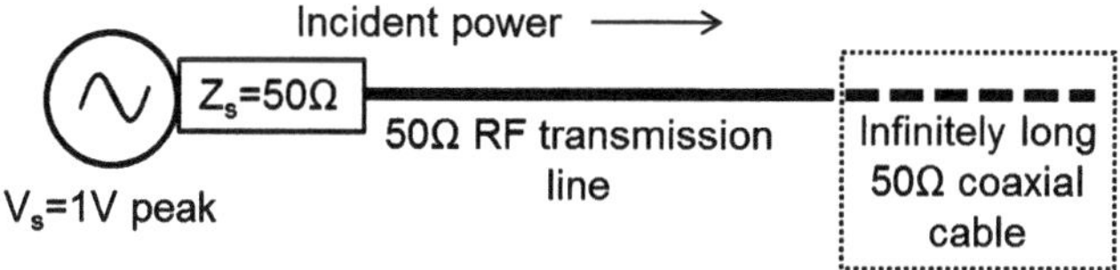

Fig. 2.14 Transmission line with 50 Ω cable

The incident voltage wave is attenuated by the cable and it does not reach the load end of the coaxial cable. Hence, there is no reflected voltage wave from the free end of the coaxial cable. It makes no difference whether the free end of the cable is open-circuited or short-circuited or connected to 50 Ω load impedance. The infinitely long coaxial cable itself behaves like a perfectly matched load. The 'infinitely' long coaxial cable concept is used to measure the VSWR characteristic of RF coaxial cables, known as Structural return loss of RF coaxial cables. The military specifications for RF coaxial connectors indicate that a RF coaxial cable is considered 'infinitely' long if the cable measures a loss of 26 dB or more at the test frequency.

2.12 VSWR: Partially Matched Load

The RF transmission line, partially matched load (less than 50 Ω or greater than 50 Ω), is shown in Fig. 2.15 for one wave length (λ) from the load end.

For load impedances greater than 50 Ω, the standing wave pattern is similar to open circuited transmission line i.e. the peak value of the standing wave always occurs at $\lambda/2$ or its multiple from the load end. For load impedances less than 50 Ω, the standing wave pattern is similar to short circuited transmission line i.e.

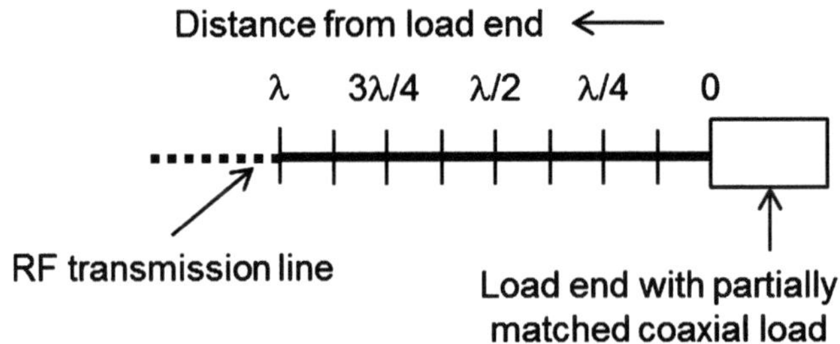

Fig. 2.15 Transmission line with partially matched load

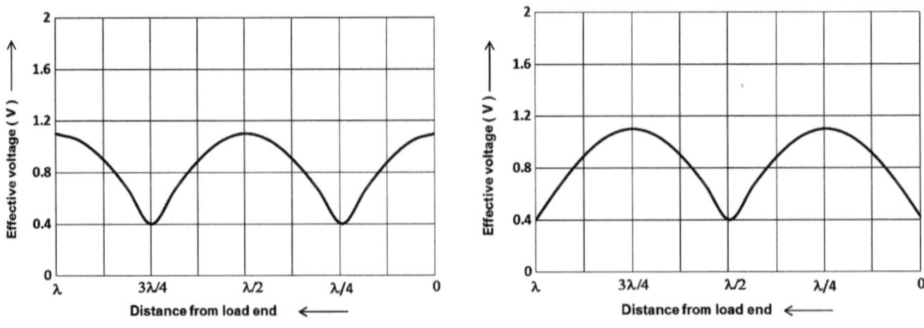

Fig. 2.16 Standing wave for Z > 50 Ω   **Fig. 2.17** Standing wave for Z < 50 Ω

the peak value of the standing wave always occurs at $\lambda/4$ or its multiple from the load end. The values of V_{max} and V_{min} depend on the deviation load impedance from 50 Ω. The effective (rms) standing voltage waveforms for the load impedances are shown in Figs. 2.16 and 2.17.

The values of V_{max} and V_{max} depend upon the amount of mismatch and other losses in transmission line. Any text book on RF transmission lines or the text book mentioned in the reference-3 could be referred for the mathematical analysis of incident and reflected waves.

References

1. Cripps SC (2008) Pulling up. IEEE Microwave Mag 9:40–48
2. Ryder JD (1965) Networks, lines and fields. Prentice-Hall, Inc., NJ
3. Matthei GL, Young L, Jones EMT (1980) Microwave filters, impedance-matching networks, and coupling structures. Artech House
4. Agilent Network Analyser Basics (2004) Agilent Technologies, Inc., 5965–7917E
5. HP Journal (1950) Greater reliability in UHF impedance measurements, Technical Information from the Hewlett-Packard Laboratories, vol 1

Chapter 3
Requirements of RF Filter

Abstract The characteristics of RF filters are Pass band, Pass band ripple, Insertion loss, Return loss, Cut-off frequency and Rejection. The characteristics are explained for the filters using customer electrical specifications with random limits and using Network analyser displays. Mechanical and environmental specifications also form the part of customer specifications in industrial environment and they are discussed in detail.

3.1 Sub-Contracting Process

Requirements of RF filters arise during the design of equipments for suppressing the unwanted band of frequencies. The normal practice of equipment designers is to sub-contract the design and manufacturing of the filters to capable auxiliary industries. As the filters have to interface satisfactorily with the circuits of equipments, the electrical, mechanical and environmental interfacing requirements of RF filters are generated by the designers of equipments in the form of specifications. The specifications provided by equipment designers (customers) are the basic inputs for suppliers to design of RF filters. The sub-contracting process for RF filters is shown in Fig. 3.1. The electrical, mechanical and environmental specifications of RF filters are explained with examples.

3.2 Sample Electrical Specifications of Filters

Sample electrical specifications for low pass, high pass, band pass and band stop filters are shown in Tables 3.1–3.4 respectively with random limiting values for the characteristics. The random values assist in better understanding of the characteristics of filters.

D. Natarajan, *A Practical Design of Lumped, Semi-Lumped and Microwave Cavity Filters*, Lecture Notes in Electrical Engineering, DOI: 10.1007/978-3-642-32861-9_3, © Springer-Verlag Berlin Heidelberg 2013

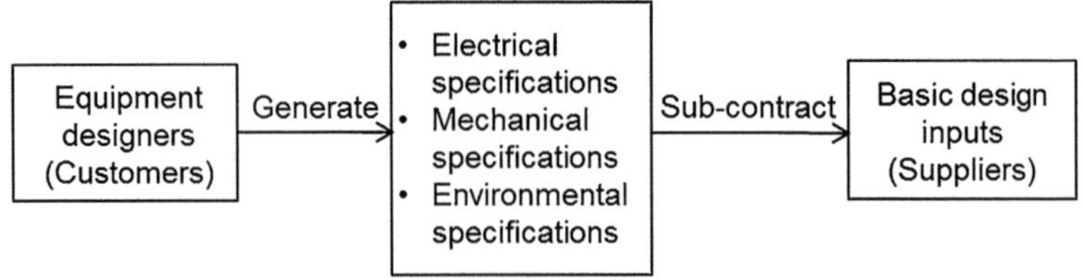

Fig. 3.1 Subcontracting process for RF filters

Table 3.1 Requirements of a low pass filter with random limits

Low pass filter	
Parameter	Limit
Pass band	(10–25) MHz
Insertion loss	1.5 dB max.
Return loss	20 dB min.
Rejection (40–500) MHz	60 dB min.

Table 3.2 Requirements of a high pass filter with random limits

High pass filter	
Parameter	Limit
Pass band	(35–100) MHz
Insertion loss	1 dB max.
Return loss	20 dB min.
Rejection (10–25) MHz	60 dB min.

Table 3.3 Requirements of a band pass filter with random limits

Band pass filter	
Parameter	Limit
Centre frequency fo	1 GHz
Pass band	fo ± 40 MHz
Insertion loss	1 dB max.
Return loss	20 dB min.
Pass band ripple	0.4 dB max.
Rejection at fo ± 100 MHz	60 dB min.

Table 3.4 Requirements of a band stop filter with random limits

Band reject filter	
Parameter	Limit
Centre frequency fo	1.5 GHz
Rejection at fo ± 50 MHz	60 dB min.
Pass band low (GHz)	1.2–1.4
Pass band high (GHz)	1.6–1.8
Pass band insertion loss	1 dB max.
Return loss	20 dB min.

Pass band, Pass band ripple, Insertion loss, Return loss, and cut-off frequency are pass band characteristics whereas rejection is stop band characteristic of filters. The method of specifying filter characteristics could vary among equipment designers. In addition to specifying pass band, 1 dB bandwidth could also be specified for band pass filters. Rejection could also be specified at 30 MHz in addition to 40 MHz for controlling the sharpness of rejection characteristic of filter.

3.3 Pass Band

Pass band of a filter is defined as the spectrum (range) of frequency that pass through the filter with acceptable or specified loss. It is characterised by a band of frequencies in KHz or MHz or GHz as applicable. For low pass filters, pass band is from DC to the specified frequency. The pass band of the low pass filter, shown in Table 3.1 is (10–25) MHz. For high pass filters, pass band begins from the specified frequency. Ideally, the pass band should stretch up to infinity. However, cost and technology limit the upper frequency of a high pass filter. The pass band of the high pass filter in Table 3.2 is (35–100) MHz.

For band pass filters, pass band is between the specified lower and upper frequencies. The pass band of the band pass filter in Table 3.3 is (960–1040) MHz. Band reject filters have two pass bands. One pass band is below the specified lower stop band frequency and the other is above the specified upper stop band frequency. The two pass bands of the band reject filter in Table 3.4 are (1.2–1.4) and (1.6–1.8) GHz.

3.4 Pass Band Ripple

Certain applications of microwave band pass filters require controlling the variation of insertion loss in the pass band of band pass filters though all the observed values of insertion loss are within limit. The variation in insertion loss is defined as pass band ripple. Mathematically, pass band ripple is the difference between the observed maximum and minimum values of insertion loss. Excessive pass band ripple generates noise in RF circuits.

For the band pass filter in Table 3.3, the insertion loss limit is 1 dB max. and pass band ripple limit is 0.4 dB max. Assume the observed insertion loss values of the filter are 0.6, 0.3, 0.2, 0.8 and 0.5 dB in the pass band. The observed values are within the specified loss and are shown in Fig. 3.2 with expanded scale for the pass band. The pass band ripple is (0.8–0.2) dB i.e. 0.6 dB, which exceeds the limit, 0.4 dB max.

Fig. 3.2 Ripple in pass band

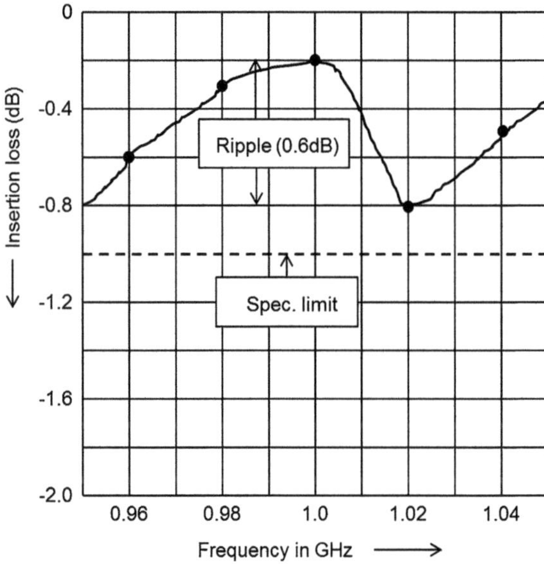

Fig. 3.3 Insertion loss of RF filter

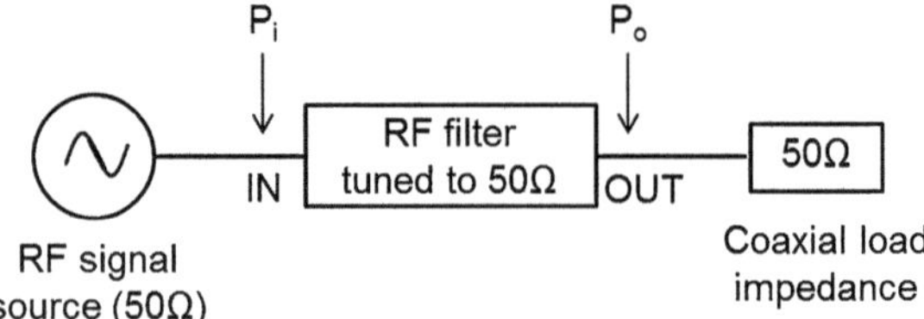

3.5 Insertion Loss

RF power is precious and it needs to be managed efficiently. Adding a RF component in a signal processing circuit attenuates power that otherwise would have reached the output of the circuit. Hence, there is a need to characterise the loss of every RF component that is inserted in RF circuits.

Figure 3.3 shows a simple RF filter circuit, terminated with load impedance. The source impedance and the load impedance are assumed to be 50 Ω. P_i is the power at the input of the filter. It is also assumed that the input and output impedance of the filter is tuned to 50 Ω.

Without the RF filter in the circuit, the power across the load impedance would be P_i. With the filter in the circuit, a fraction of the power, P_i, is dissipated in the filter and the balance power, P_o, reaches the load impedance. The dissipated power is known as the insertion loss of filter. Insertion loss of a RF filter is defined as the additional attenuation or loss caused by the filter in the pass band. As the insertion loss results in transmission loss, it is a transmission parameter. The insertion loss is expressed in dB.

$$\text{Insertion loss(dB)} = 10 \log(P_i) - 10 \log(P_o) = 10 \log(P_i/P_o)$$

As $P_i > P_o$, insertion loss computed by the expression has a positive value. For example, assume P_i is 10 mW and only 9 mW reaches the load due to filter loss.

$$\text{Insertion loss} = 10 \log 10 - 10 \log 9 = (10 - 9.54)\,\text{dB} = 0.46\,\text{dB}$$

Network analyser is used for measuring the insertion loss of filters. It displays the insertion loss with a negative sign. It has precision source and detector (load) ports. The device under test is connected between the ports and the power level at the detector port is measured in dB. Initially, the power, P_i, is first directly applied to the detector port and the power level in dB i.e. 10 log (P_i) is set to zero for convenience. If a filter or any other RF component is inserted between the two ports, it displays the reduction in power level caused by the RF component in dB from the initial reference power level. The reduction in power level is the insertion loss of the component and it is displayed in dB with a negative sign. For example, if $P_i = 10$ dB (10 mW) is applied to the detector port, it is set to zero by subtracting 10 dB. With a filter connected between the ports, the power level at the detector port is 9.54 dB (9 mW) but it is displayed as (9.54–10) dB i.e. −0.46 dB, which is the insertion of the filter.

The insertion loss filters should be met throughout the pass band. The insertion loss of filters is not just dissipated power alone. The filters cannot be tuned to 50 Ω exactly and they could only be tuned close to 50 Ω. Hence, filters cause some amount of mismatch loss, which is added to insertion loss.

3.6   Return Loss

Figure 3.4 shows a simple RF filter circuit, terminated with load impedance. The source impedance and the load impedance are assumed to be 50 Ω. V_{in} is the voltage at the input of the filter. It is assumed that the input and output impedance of the filter is not tuned to 50 Ω. Ignoring the mismatch caused by the RF source and the load, the primary source of mismatch is the RF filter in the circuit. Hence, a voltage, V_{ref}, is reflected by the filter towards the source due to mismatch.

Reflection coefficient $(K) = V_{ref}/V_{in}$

$$\text{VSWR} = \frac{(1 + |K|)}{(1 - |K|)}$$

Filters are tuned to lowest VSWR in the pass band to minimise mismatch loss, which otherwise would add to the insertion loss of the filters. Typically, a VSWR of 1.22 maximum is acceptable for filters in most applications.

Network Analyser is used to tune the filter for complying with the specified requirements of customers. Tuning is accomplished by having frequency in x-axis and attenuation in dB in y-axis in the display of the Network Analyser so that all the characteristics of a filter could be viewed simultaneously. All the

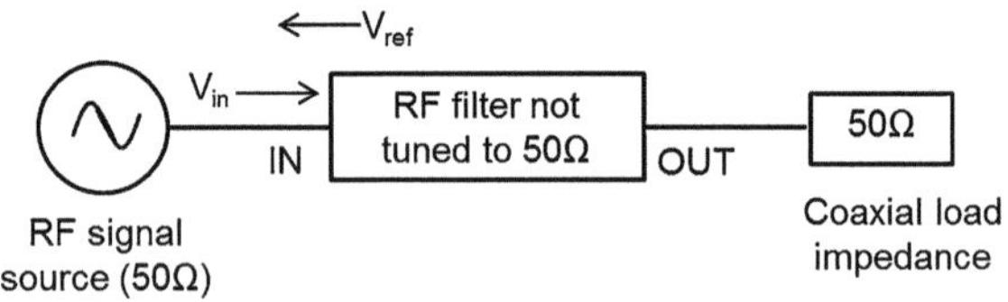

Fig. 3.4 Reflected voltage in mismatched RF filter

characteristics of a filter are expressed in dB except VSWR, which has no units. Hence, VSWR is converted into another parameter known as, Return loss, which is expressed in dB. Return loss is related to VSWR by the expression given below.

$$\text{Return loss(dB)} = -20 \log \left[\frac{\text{VSWR} - 1}{\text{VSWR} + 1} \right]$$

VSWR of 1.22 corresponds to a return loss of -20 dB approximately. Return loss is specified as -20 dB maximum or 20 dB minimum. Return loss should be met throughout the pass band of filters. As return loss results in reflection losses, it is classified as a reflection parameter. Unless otherwise specified by customers, return loss is ensured only at the input of filters. The input and output of filters is identified as 'IN' and 'OUT' by filter manufacturers.

3.7 Cut-Off Frequency

After the pass band, the insertion loss of filters increases. The frequency at which it becomes 3 dB is called cut-off frequency. Very few customers indicate cut-off frequency by specifying 3 dB band width. An example could be given for a band pass filter.

$$\text{fo, Centre frequency} : 1,000 \, \text{MHz}$$

$$\text{Pass Band} : (\text{fo} \pm 40) \, \text{MHz}$$

$$3 \text{dB Bandwidth} : 250 \, \text{MHz min}$$

The cut-off frequencies of the band pass filter are calculated and they are shown in Fig. 3.5, expanding the pass band of the filter.

$$\text{Lower cut-off frequency} = (\text{fo} - 125) = 875 \, \text{MHz}$$

$$\text{Upper cut-off frequency} = (\text{fo} + 125) = 1125 \, \text{MHz}$$

Pass band needs to be extended for satisfying 3 dB bandwidth. Extending bandwidth improves insertion loss in the pass band but additional efforts are required for complying with rejection requirements.

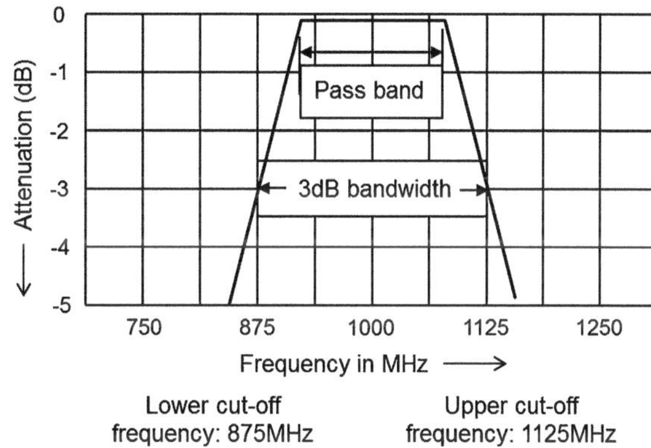

Fig. 3.5 Cut-off frequencies of band pass filter

3.8 Rejection

Rejection characteristic of a filter is defined as the capability of a filter to attenuate the unwanted band of frequencies and it is expressed in dB. Rejection characteristic could be considered similar to the insertion loss of a filter in the sense that the filter offers very high insertion loss for the frequencies in the rejection band. For example, the rejection of the band pass filter in Table 3.3 is specified as 60 dB min. at $(f_o \pm 100)$ MHz. It could be interpreted as the insertion loss of the band pass filter is 60 dB min. in the frequencies $(f_o + 100)$ and $(fo - 100)$ MHz.

Referring to Fig. 3.3, the power, Po, across the load is 60 dB down from the input power, Pi, at the rejection frequencies i.e. the filter has practically attenuated or rejected these frequencies from reaching the load. Like insertion loss, rejection is also specified with positive sign or negative sign. For example, it could be specified as 60 dB min. or −60 dB max. Network analyser displays rejection with a negative sign.

3.9 Graphical Representations

The electrical characteristics of RF filters are presented graphically using the typical acceptable characteristics. The performance graphs for the four types of filters in Tables 3.1–3.4 are shown in Figs. 3.6–3.9 respectively. The graphs are actually the displays of Network analyser showing all the performance characteristics of tuned filters. Frequency is in X-axis and attenuation is in Y-axis. 0 dB is the reference for attenuation.

Fig. 3.6 Typical
characteristics of low pass
filter

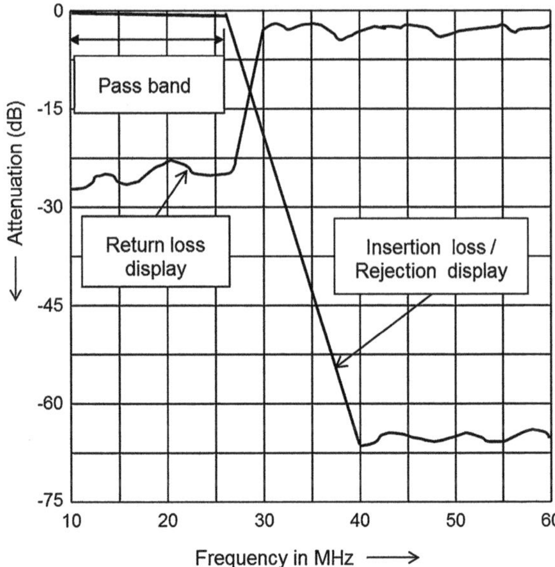

Fig. 3.7 Typical
characteristics of high pass
filter

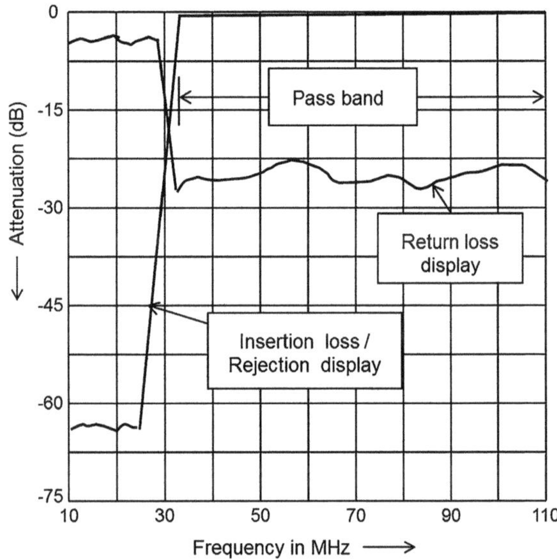

3.10 Mechanical Specifications

Mechanical specifications generally include dimensions (outline), weight, input/
output connections and the external finish of filters. More information about the
mechanical specifications is provided below.

Fig. 3.8 Typical characteristics of band pass filter

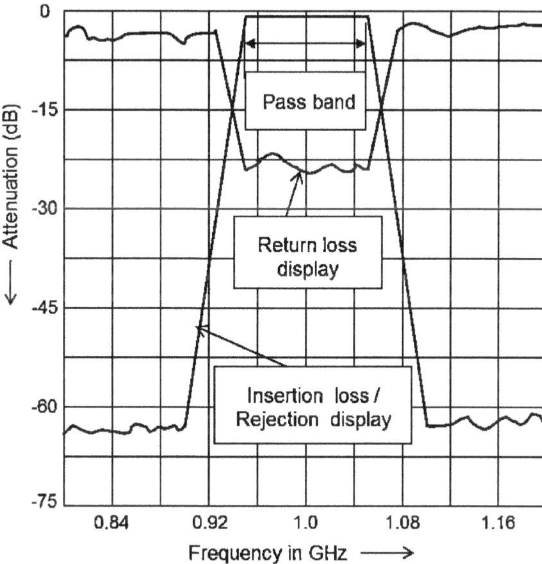

Fig. 3.9 Typical characteristics of band reject filter

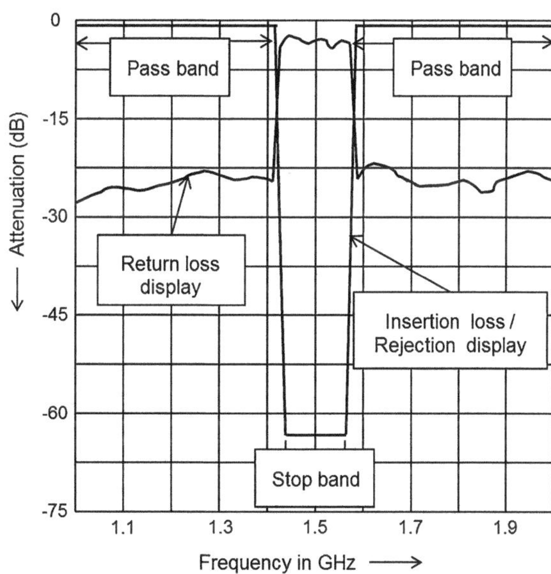

3.10.1 Dimensions (Outline) and Weight

Dimensions (outline) and weight are generally flexible for new designs of RF filters and they are not specified by customers for most filters. Filter designers indicate them to customers at the end of the development of filters. However, they are controlled by customers if the filters are second sourced or are used in man-pack sets.

Dimensions (outline) refer to the overall length, width and height excluding the input/output connections and the projection of tuning and coupling screws, if any. Filter designers usually limit the projection of tuning screws approximately two to three times the thickness of lock nuts in microwave cavity filters. Usually blind threaded holes are used for mounting filters in equipments. The location and the size of blind threaded holes for the mounting of filters are also part of dimensions (outline).

3.10.2 Input/Output Connections

RF coaxial connectors or cylindrical tab terminations are used for connecting the input/output of filters. SMA, BNC and TNC coaxial connectors are generally preferred. Cylindrical tab terminations are used for the surface mounting of filters and they are cheaper compared to coaxial connectors. Additional design information about the tab terminations is explained in Chap. 5, Planning the design of RF Filters. Dimensions of input/output coaxial connectors need not be shown. If the dimensions of the coaxial connectors need to be shown, they should be governed by the international standards applicable to the connector series.

3.10.3 Finish

Metal enclosures are generally used for filters. Conductive chromate conversion or silver plating is done externally for the enclosures depending on the frequency of application of filters. Suitable additional finish such as painting or lacquer coating could also be specified for the external surface of the enclosures.

3.11 Environmental Specifications

Filters are generally mounted inside systems. Specifying operating temperature, and humidity limits are adequate. Water sealing requirement is specified for filters used in out-door applications such as those mounted on transmission towers. An example of specifying environmental specifications is shown below and it could be applied for most applications.

$$\text{Operating temperature} : 0 - 50^{\circ}\text{C}$$

$$\text{Humidity} : 40^{\circ}\text{C}, 95\,\%\text{RH}$$

Sealing (when applicable): Rain/Water immersion/Dust tests, as applicable.

For military applications, the specified operating temperature could be from −40 °C to +85 °C and additional tests such as salt spray may be included. It is quite common to link the environmental tests to established Standards for controlling the method of conducting the environmental tests.

3.12 Additional Specifications

Materials that are used in filters are not normally specified by equipment designers. However, filters used in medical electronic equipments must be non-magnetic to avoid erroneous diagnosis. Non-magnetic requirement is specified by customers indicating the magnetic field intensity (Ex.: 0.5 or 2 T) in which the filters have to perform. Such a requirement forces filter designers to select appropriate materials such as non-magnetic brass or phosphor bronze or other non-magnetic materials for all the parts of filters to satisfy the magnetic field intensity requirements.

Chapter 4
RF Filter Design Technologies

Abstract The design of RF filters is complex compared to the design of power frequency filters used at the output of rectifier circuits to smooth ac ripple components. Many technologies are available considering frequency, performance, size, power handling and cost of RF filters. The L-C circuit configurations for lumped RF filters are presented with the functioning of the circuits. The concept of semi-lumped RF filter, resonance in microwave coaxial cavity, combline and iris-coupled cavity filters are technologies are explained in detail. Information regarding power handling capacity, equivalent circuit and factors to minimise insertion loss of cavity filters are provided. The basic microstrip and stripline circuit and the computer aids for designing the filters are also explained.

4.1 Technologies

The basic function of RF filter is to pass required band of frequencies with minimum loss and to attenuate unwanted frequency components to the desired levels. The simplest filters are those used at the output of half wave or full wave rectifier circuits to pass smooth dc output voltage by filtering out ac ripple components. The power frequency filters use large sized inductors and electrolytic capacitors in the form of L or Π sections and they are designed using standard plots [1] to limit ripple voltage to desired level. As the frequency increases, the design of RF filters becomes complex and many technologies are available to design the filters for varied application needs. Frequency, performance, size, power handling and cost decide the appropriate technology for designing RF filters.

Application of filters might emphasize a particular characteristic. Some application requires more flat pass band characteristic with minimum pass band ripple and such filters are classified as maximally flat type [2] or Butterworth type. Some application requires sharp fall from pass band limiting frequency for

D. Natarajan, *A Practical Design of Lumped, Semi-Lumped and Microwave Cavity Filters*, Lecture Notes in Electrical Engineering,
DOI: 10.1007/978-3-642-32861-9_4, © Springer-Verlag Berlin Heidelberg 2013

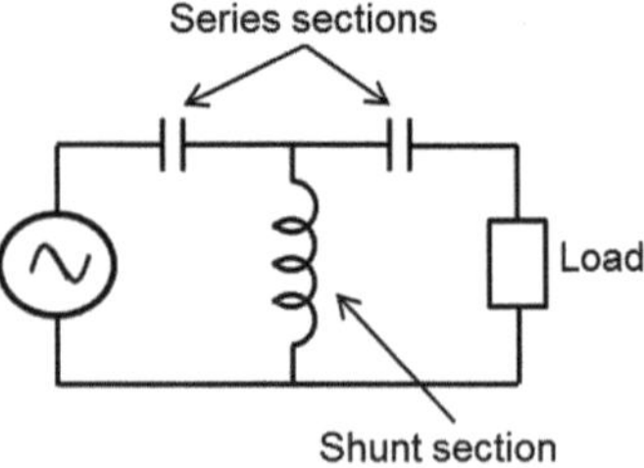

Fig. 4.1 Series and shunt sections of filter circuit

realising high rejection characteristic and such filters are classified as Tchebyscheff type [2]. Three technologies that are available to design RF filters for application needs are explained. The technologies are:

1) Lumped and Semi-lumped RF filters.
2) Microwave cavity filters.
3) Microstrip and Stripline filters.

4.2 Lumped RF Filters

RF filters that use discrete inductors and capacitors are called lumped RF filters. A hand or machine wound inductor is an example of discrete inductor. A chip ceramic capacitor is an example of discrete capacitor. Lumped filter technology is used for designing low pass, high pass, band pass and band reject filters. With the advancements in chip ceramic capacitors, lumped RF filters are designed up to 100 MHz or higher.

In lumped RF filters, inductors and capacitors are networked in the form of series and shunt sections. The sections which are in the path of signal source and load are called series sections. The sections which are across the signal source or load are called shunt sections. Filters are designed in two configurations of LC circuit, namely T-section and Π-section. T-section filter has a series section at its input and Π-section filter has a shunt section at its input. The filter circuit shown in Fig. 4.1 is a T-section filter with two series and one shunt sections.

Both T-section and Π-section configurations are explained for the four types of filters. The number of sections in the filters is limited to three for explaining the four types of filters.

4.2.1 Low Pass and High Pass Filters

The T-section and Π-section LC circuit configurations for low pass filter are shown in Fig. 4.2 and the same for high pass filters are shown in Fig. 4.3.

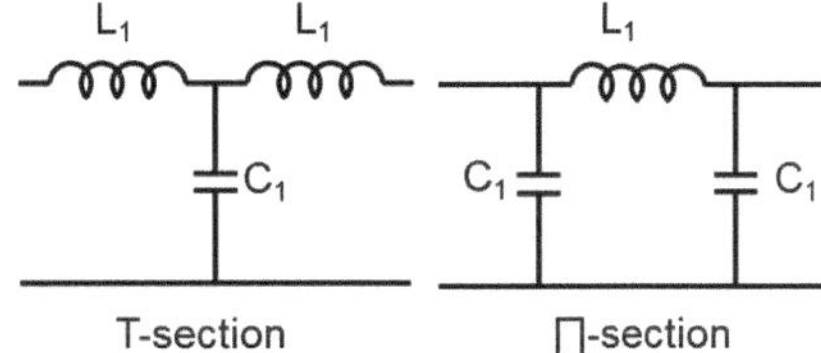

Fig. 4.2 Configurations of low pass filter

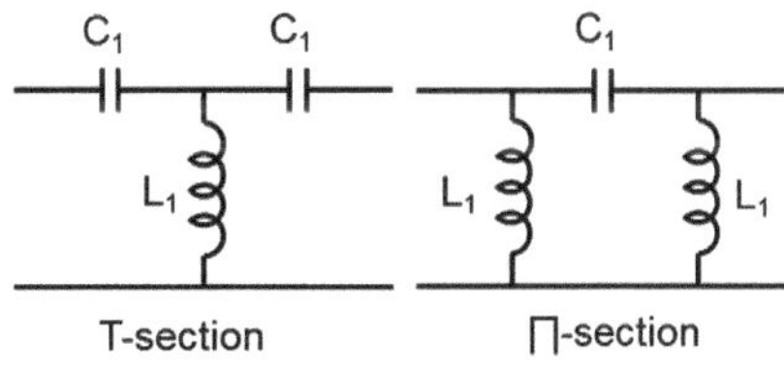

Fig. 4.3 Configurations of high pass filter

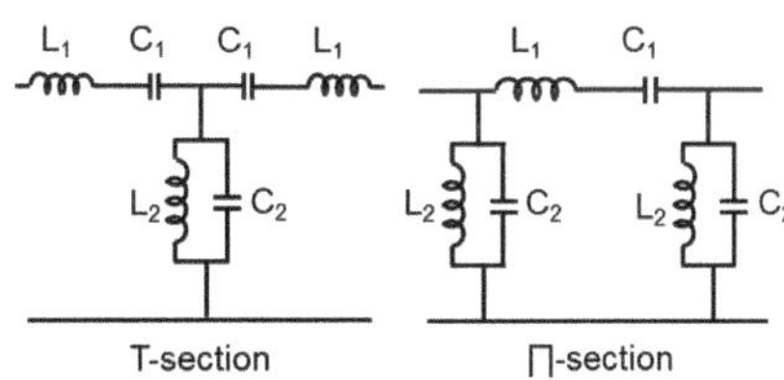

Fig. 4.4 Configurations of band pass filter

In low pass filter, the reactance of series inductive elements increases and that of shunt capacitive elements decreases at frequencies above the upper pass band edge frequency of the filter, causing the rejection. In high pass filter, the reactance of shunt inductive elements decreases and that of series capacitive elements increases at frequencies below the lower pass band edge frequency of the filter, causing the rejection.

Pass band insertion loss of filters is decided by:

1) The quality factors (loss factors) of the reactive components:

Purity of copper, length/diameter of copper wire and finish on the wire contributes the quality factor of inductors. The loss factor of capacitors is characterised by $\tan \delta$ specified by the manufacturers of capacitors. Detailed guidance for the selection of inductors and capacitors is provided in Chap. 6.

2) The layout of the inductors and capacitors also influences pass band insertion loss to a large extent.

4.2.2 Band Pass and Band Reject Filters

The T-section and Π-section LC circuit configurations for band pass are shown in Fig. 4.4 and the same for band reject filters are shown in Fig. 4.5.

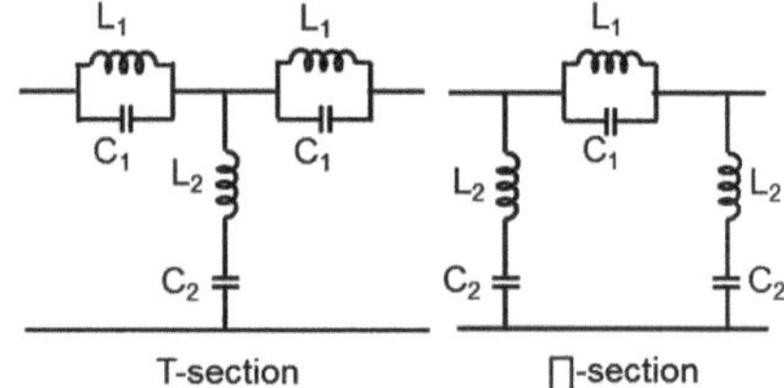

Fig. 4.5 Configurations of band reject filter

In band pass and band reject filters, the series and shunt sections resonate at the centre frequency of pass band or stop band. In band pass filters, the series sections offer minimum resistance and the shunt sections offer maximum resistance at the centre frequency of the pass band. Hence, the frequency components within the pass band are passed with minimum loss and the frequency components outside the pass band are rejected with high attenuation.

In band stop filters, the series sections offer maximum resistance and the shunt sections offer minimum resistance at the centre frequency of the stop band. Hence, the frequency components within the stop band are rejected with high attenuation and the frequency components outside the pass band are passed with minimum loss. Pass band insertion loss of the filters is decided by the quality factors (loss factors) of the reactive elements and the layout of the elements. More reactive sections are added to increase the rejection capability of band pass and band reject filters. However, adding more reactive sections increases the insertion loss of filters.

The design of lumped RF filters involves calculating the values of inductors and capacitors of filters for the specified band width. The design of lumped low pass filter, Tchebyscheff type, is explained in tutorial form illustrated with an example in Chap. 6, Design of lumped and semi-lumped RF filters. Additional design information on the selection of inductors and capacitors and the layout of the elements is also provided.

4.3 Semi-Lumped RF Filters

RF filters that use discrete inductors and distributed capacitors are called semi-lumped RF filters. Distributed form of a capacitor means that the required capacitance is developed in the design of the filters instead of using a discrete capacitor. For example, coaxial connectors are used as input and output connectors for filters. The capacitance between the centre and outer conductors of the connectors is an example of distributed capacitor. A more realistic example of the distributed capacitor is explained.

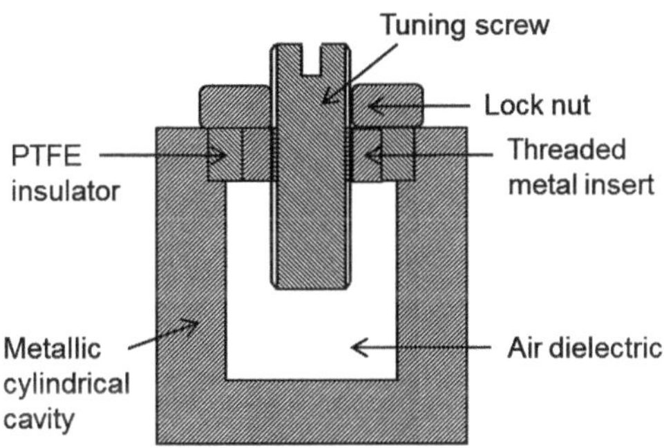

Fig. 4.6 Variable distributed capacitor in cavity

4.3.1 Distributed Capacitance

Figure 4.6 shows a mechanical sub-assembly which develops distributed capacitance. The sub-assembly consists of a PTFE insulator and a knurled metal insert having internal threads, press-fitted on to a metallic cylindrical cavity. A tuning screw with its lock nut is inserted into the cavity through the internal threads of the knurled metal insert.

Distributed capacitance exists between the tuning screw and the metallic cylindrical cavity in the assembly. The PTFE insulator and the air inside the cavity act as dielectric developing distributed capacitance. The value of the capacitance could be slightly adjusted by varying the longitudinal position of the tuning screw. The value of the distributed capacitance depends upon the dimensions of the cavity and tuning screw. The dimensions and the dielectric constant of the PTFE insulator also contribute for the distributed capacitance. For example, the use of larger diameter tuning screw develops higher distributed capacitance for the same cavity dimensions. The design expression for computing the distributed capacitance is available in Sect. 6.3.3, Design of lumped and semi-lumped RF filters.

4.3.2 Concept of Semi-Lumped RF Filter

A mechanical design could be developed for electrically connecting two discrete inductors to the knurled metal insert as shown in Fig. 4.7a.

It could be visualized that the mechanical arrangement is actually a T-section semi-lumped low pass filter and the equivalent circuit of the filter is shown in Fig. 4.7b.

As the values of distributed capacitors that could be realised are small, semi-lumped RF filters are designed for high frequency applications, ranging from

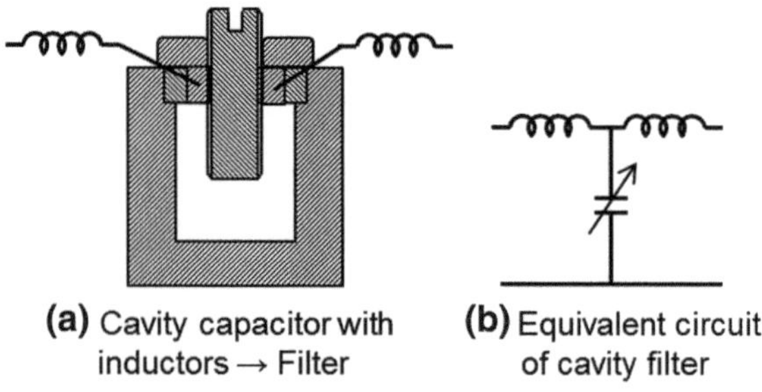

Fig. 4.7 Example of semi-lumped low pass filter

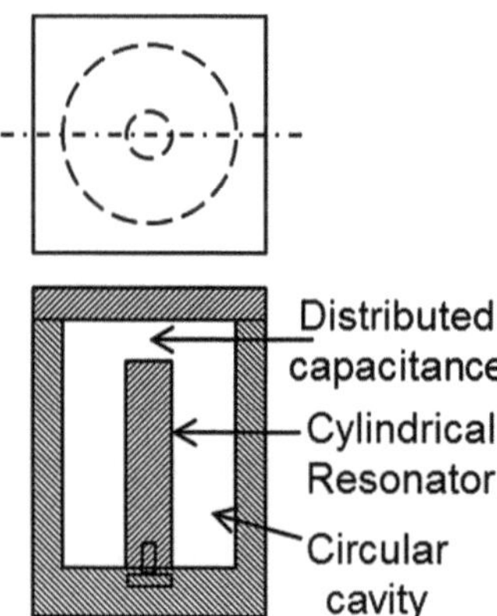

Fig. 4.8 Microwave cavity with a coaxial resonator

100 to 500 MHz or higher. The insertion loss of semi-lumped RF filters are lower than their equivalent lumped RF filters as the Q-factors of distributed capacitors are higher than those of discrete capacitors.

Semi-lumped RF filter technology is used to design all the four types of filters. It is relatively easier to realise low pass and band pass filters in tubular form. The compact structure of a tubular filter and wide relative bandwidth are attractive for many applications. The design of semi-lumped low pass filter, Tchebyscheff type, is explained in tutorial form with an example in Chap. 6.

4.4 Microwave Cavity Filters

4.4.1 Resonance in Microwave Coaxial Cavity

A microwave coaxial cavity is a metallic enclosure in the form a cylinder with a resonator, a cylindrical element, mounted coaxially in the enclosure. The resonator is mounted in the cavity with one end and the other end of the resonator is free. The mechanical arrangement of a microwave coaxial cavity is shown in Fig. 4.8.

Cavities having rectangular or square cross section could also be used. Distributed capacitance is developed at the free end of the resonator. The inductance of the resonator and the distributed capacitance resonate at a frequency decided by the mechanical dimensions of cavity and resonator. The resonant section is a shunt section. The microwave coaxial cavity with resonator is the basic resonant section in microwave cavity filters. The number of resonant sections in the filters is decided by the electrical specifications of filters. The propagation of electromagnetic waves in cavity filters is TEM mode. Inductive coupling exists between adjacent resonant cavity sections. End resonant cavity sections are coupled to input/output coaxial connectors. Combline and iris-coupled microwave cavity filters are quite popular in industrial applications. Combline filters use $\lambda/8$ resonators and iris-coupled filters use $\lambda/4$ resonators at resonant frequency.

4.4.2 Combline and Iris-Coupled Cavity Filters

In microwave cavity filters, air is the dielectric medium of transmission and the filters do not require any dielectric support. The filters are characterised by low insertion loss with high rejection making them suitable for high power applications. A cavity width of 30 mm could easily handle 100 W of RF power.

Microwave cavity filter technology is explained for the two most commonly used types of band pass filters, namely:

1) Combline band pass filter.
2) Iris-coupled band pass filter.

Combline cavity technology is applied to design microwave band pass filters having a bandwidth up to 15 % of f_o, centre frequency [2]. Iris-coupled cavity technology is applied to design microwave band pass filters having a narrow bandwidth of less than 1 % of f_o, centre frequency [2].

Combline microwave filters are used extensively in mobile and satellite communication systems in recent years because of their compact size, low cost, wide tuning range, relatively low loss and good spurious response [3–5]. Iris-coupled band pass filters are widely used in wireless base-stations [3].

Band pass filters that are designed with combline cavity structure have several advantages:

1) Combline cavity filters are very compact.
2) It is easier to realise high rejection for the stop bands.
3) Milling of combline cavity block is relatively much faster and cheaper compared to iris-coupled cavity band pass filters.
4) Combline filters are relatively easier to assemble ensuring faster production.

Iris-coupled band pass filter have the advantage of achieving narrow band width and lower insertion loss although the milling operation of cavities is expensive compared to combline cavities.

4.4.3 Power Handling Capacity

Power handling capacity of RF filters needs to be established when the filters are used in high power transmitters or at high altitude conditions. Multipaction breakdown could occur in filters used under high altitude conditions. Multipaction is a discharge of electrons due to low pressure that is associated with high altitudes. Waveguides are pressurised as per the needs of high altitudes to over come the breakdown.

Creep distance between filter elements needs to be examined for filters used in high power applications. Ionisation breakdown could occur if the creep distance is marginal at the power level in which the filters are used. Frequency of operation should also be considered in deciding the creep distance between various filter elements.

Intermodulation distortion occurs in RF filters used at very high power applications. Intermodulation products are due to poor or variation in contact resistance between filter parts. For example, poor contact between input/output coaxial connectors and the cavity block of filter results in intermodulation distortion. Appropriate mechanical design for the assembly of filter parts is necessary to prevent the generation of intermodulation products. Soldering of parts, where feasible, minimises intermodulation distortion considerably.

RF filters dissipate RF power in all applications but the dissipated power might become significant for filters used in very high power applications. The dissipated power causes the surface temperature of RF filters to increase. A rise of 15 to 20 °C above the ambient temperature is acceptable. Higher temperature rise affects the electrical performance of filters and causes the degradation of dielectric material in substrate filters. Additional considerations are required for high power filters used at high altitude conditions [6].

4.5 Combline Band Pass Filters

4.5.1 Structure

The structure of combline cavity band pass filter with its side cover plate removed is shown in Fig. 4.9.

The filter has three important piece parts, namely, cavity block, resonator line or simply known as resonators and cover plate. Though the cavity block is physically one cavity, it should be considered as a series of coupled cavities, each having a resonator. In Fig. 4.9, there are three sectional cavities and each sectional cavity has a brass resonator. The resonators are mounted in the cavity with one end grounded i.e. short circuited to the cavity. Distributed capacitance with air as dielectric is developed at the free end of the resonators. The value of the distributed capacitance is adjusted with screws. The cover plate is mounted on cavity block to form closed sectional cavities.

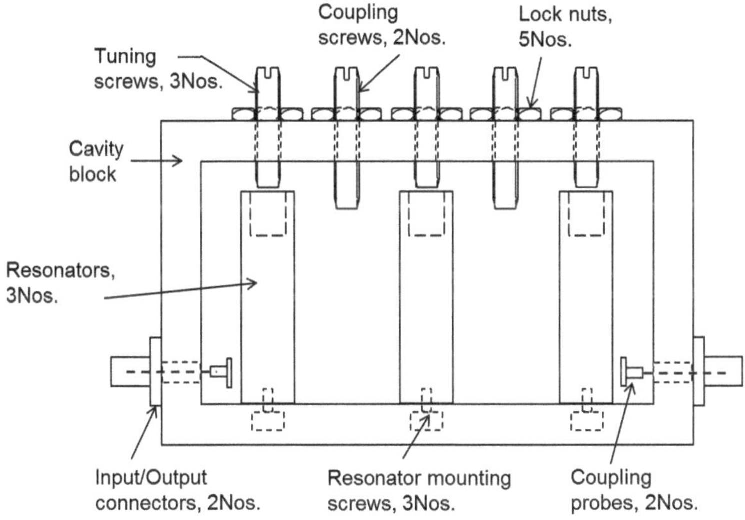

Fig. 4.9 Structure of combline band pass filter (Without side plate)

The resonators are $\lambda/8$ long at the centre frequency (f_o) of band pass filter but shorter lengths could also be used. The resonators are mounted perpendicular to the length of the cavity block with designed spacing and they are centred to the width of the cavity. Each sectional cavity is tuned to the centre frequency, f_o, of band pass filter using its resonator and distributed capacitance. Two coupling screws are shown in Fig. 4.9 for the three section combline filter. The coupling screws adjust the capacitive coupling between the sectional cavity resonators. Input and output coaxial connectors are coupled to end resonators using coupling probes or tags. The number of filter sections i.e. the number of resonators are increased to enhance the rejection capability filters but it increases the insertion loss also.

4.5.2 Equivalent Circuit

The equivalent circuit [7] of a combline band pass filter with three resonators is shown in Fig. 4.10. The resonator inductance and the distributed capacitance form the shunt sections of the band pass filter.

Though both inductive and capacitance couplings exist between resonator sections, only an inductance is shown between the shunt sections as the predominant coupling between resonator sections is inductive [2]. Inductive coupling is maximum at the short-circuited end of resonators and the capacitive coupling is maximum at the free end of resonators. The nature of coupling between resonator sections is expressed as,

Coupling between resonators = Inductive coupling − Capacitive coupling

Fig. 4.10 Equivalent circuit
of combline filter

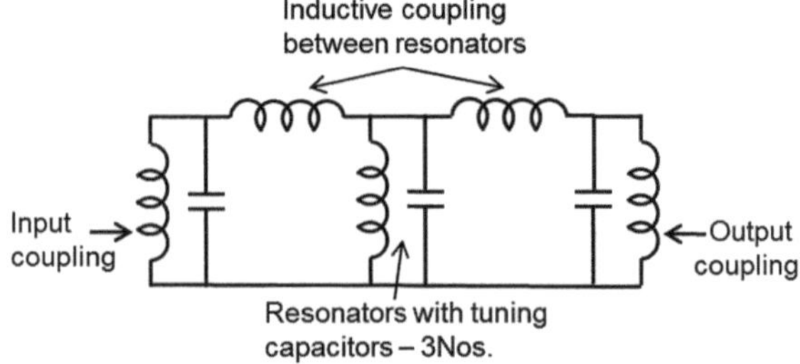

With the advancement of coupling screws into the cavity, the capacitive
coupling decreases, causing the increase of coupling between resonator sections.
Increasing the coupling between resonator sections increases the bandwidth of
combline filter.

4.5.3 Insertion Loss

Design efforts are made to minimise the insertion loss of combline filters by
maximising the unloaded Q-factor of resonators. The unloaded Q-factor of a
resonator is analogous to the Q-factor of a discrete inductor. Transmission
and reflection types [8, 9] of measurements could be used to determine the
unloaded Q-factor of resonators. The Q-factor depends on the material and finish
of resonators. Silver plating of cavity block and resonators improves the Q-factor
of the filters. Cavity filters are designed for an impedance of 70 Ω as it gives the
best Q-value for the filters [10]. Increasing the size of filter cavity also reduces the
insertion loss of filters.

The design of combline filter involves calculating the number of resonators,
capacitance for the resonators and the mechanical dimensions of cavity block,
resonators, spacing between resonators , tuning and coupling screws for the
specified centre frequency, bandwidth and rejection requirements of the filter. The
design of combline band pass filter, Tchebyscheff type, is explained in tutorial
form illustrated with an example in Chap. 7.

4.6 Iris-Coupled Band Pass Filters

The structure of iris-coupled filter with its top cover plate removed is shown in
Fig. 4.11.

Iris-coupled filter also has Cavity block, brass resonators, Cover plate, tuning
screws and coupling screws. Unlike combline filter, each resonator has a clearly
defined coaxial circular or square cavity with corners rounded-off. Each cavity is
coupled to its adjacent cavity through a narrow rectangular opening called iris.

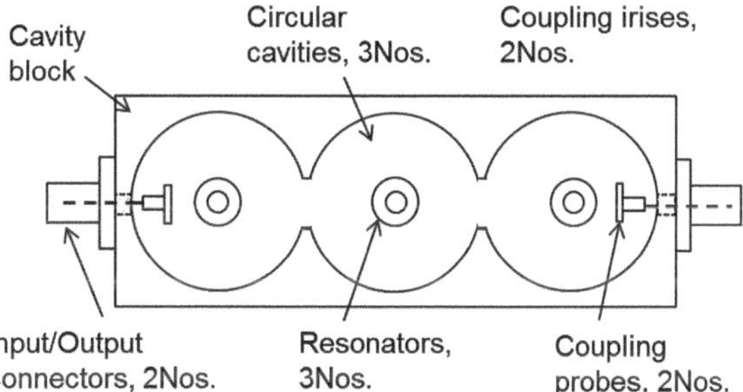

Fig. 4.11 Structure of iris-coupled band pass filter (Without top plate, and coupling screws)

The filter shown in Fig. 4.11 has three circular cavities and two coupling irises. The resonators are usually $\lambda/4$ long at the centre frequency (f_o) of band pass filter. The resonators are mounted in the cavity with one end grounded i.e. short circuited to the cavity. Distributed capacitance with air as dielectric is developed at the free end of the resonators. The value of the distributed capacitance is adjusted with tuning screws. Each cavity is tuned to the centre frequency, f_o, of iris-coupled band pass filter using its resonator and distributed capacitance. Coupling between the cavities is adjusted using coupling screws, which advance into the irises. Input and output coaxial connectors are coupled to end resonators using coupling probes or tags.

The equivalent circuit, the principles of coupling between cavity sections and minimising insertion loss of iris-coupled filter are similar to those explained for combline filter. The design of combline filter involves calculating the number of resonator sections, tuning capacitance for resonators and the mechanical dimensions of cavity block, resonators, the dimensions of irises, tuning and coupling screws for the specified centre frequency, bandwidth and rejection requirements of the filter. The design of iris-coupled band pass filter, Tchebyscheff type, is explained in tutorial form illustrated with an example in Chap. 7.

4.7 Cavity Filters with Dielectric Resonators

Microwave cavity filters could also be designed with dielectric resonators replacing brass resonators for improving the insertion loss performance of the filters. The Q-factor of dielectric resonators is very high compared to brass resonators. Dielectric resonators with unloaded Q of more than 50,000 at 18 GHz are available [11]. The size of filters increases with the increase in the Q values of dielectric resonators similar to the case of metallic resonators. For the same operating frequency band, the insertion loss of dielectric resonator filters is lower than that of brass resonator filters but the size of the filters are larger [12]. Filters with dielectric resonators have stable electrical performance over wider operating temperature range and have become popular in base stations for mobile and hand-set applications [13].

4.8 Microstrip and Stripline Filters

Substrate filter is generic name for microstrip and stripline filters. Although substrate filters have higher losses compared to cavity filters, they are widely used as they are highly compact in size. The designed filter pattern is printed on a substrate, which is a dielectric material. Details of the substrate materials and the characteristics of basic microstrip and stripline are explained.

4.8.1 Materials

Double sided glass epoxy copper clad laminates, grade FR-4 (Flame Retardant-4) or double sided PTFE copper clad laminates are used for constructing a microstrip transmission line. The sheets are available in different thicknesses. The dielectric constant of glass epoxy laminates is 4.5 approximately and it is not closely controlled. Its loss factor (tan δ) is also high in the order of 0.015. These limitations are acceptable for frequency applications up to few GHz. For higher frequencies, PTFE laminates are used as their dielectric constant is typically controlled to within $\pm$ 0.02 and the loss factor is less than 0.002. PTFE laminates are available with different dielectric constants. The designed microstrip/stripline filter circuit pattern is chemically etched on the laminates with precise control on the dimensions of conductor patterns. Ceramic substrates are also available for special applications.

4.8.2 Basic Microstrip

The basic construction of a microstip transmission line is shown in Fig. 4.12. The expressions for determining characteristic impedance, Zo, are [14]:

$$Z_0 = \frac{60}{\sqrt{\varepsilon_r}} \ln\left(\frac{8H}{W} + \frac{0.25W}{H}\right) \text{ohms for } (W/H) < 1$$

$$Z_0 = \frac{120\pi}{\sqrt{\varepsilon_r}\left[\frac{W}{H} + 1.393 + \frac{2}{3}\ln\left(\frac{W}{H} + 1.444\right)\right]} \text{Ohms for } (W/H) \geq 1$$

Microstrip antennas are one of the most common applications. Filters, attenuators, dividers, combiners and Filters are some of the other applications. The finished product is mounted inside a metal enclosure with adequate air gap between the microstrip and the top plate of the enclosure to avoid interaction between substrate and enclosure. Solder mask coated microstrip could be used for low frequency applications.

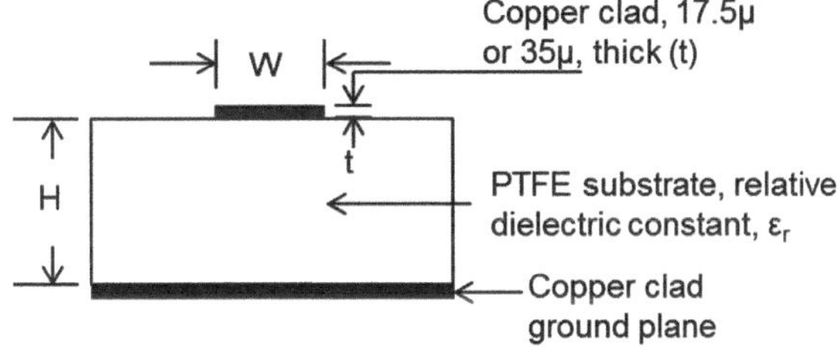

Fig. 4.12 Basic construction of microstrip transmission line

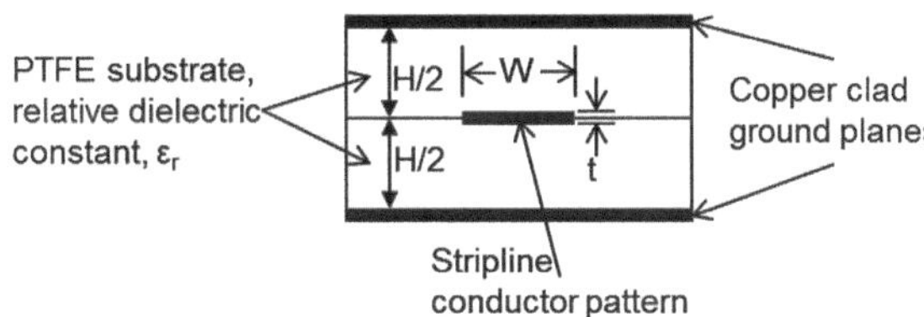

Fig. 4.13 Basic construction of stripline transmission line

4.8.3 Basic Stripline

Stripline is fabricated similar to microstrip and second substrate with ground plane is placed over it. The construction of basic stripline circuit is shown in Fig. 4.13.

The expression for determining characteristic impedance, Zo, is [15]:

$$Z_0 = \frac{60}{\sqrt{\varepsilon_r}} \ln\left[\frac{1.9H}{0.8W(0.8W + t)}\right] \text{ ohms}$$

Examples of the application of stripline technology are filters and couplers. Like microstrip circuits, stripline circuits are also enclosed in metal enclosures.

4.8.4 Computer Aids for Filter Design

Proto-model of filter is needed for verifying compliance to customer specifications. If the filter does not meet the specifications, the application of 'cut & try' [2] technique is applied or another proto-model of the filter is fabricated. It is easier to apply the 'cut & try' technique in cavity filters as it requires just filing or milling. However, the application of cut & try technique for substrate filters is practically impossible. Another proto-model of substrate filter needs to be fabricated and the fabrication involves re-fabricating substrate with its printed pattern. The process of fabricating substrate with its printed pattern is expensive and time consuming in substrate filters. Hence, the current trend is to go in for three dimensional full-wave electromagnetic (EM) simulators and analyse the performance for optimum results considerably reducing design cycle time.

Three dimensional electromagnetic field simulators are available commercially for designing RF filters. The simulators run on desktop systems and enable the development and prediction of performance of the filters without the need to fabricate them [16]. If the desired performance of the filters is not achieved, the filters are re-designed using the simulators. Design cycle time of the filters is reduced considerably with lower time-to-market [17]. As the simulators support many technologies, one can design and simulate the performance of filters with appropriate RF knowledge and training.

Microwave literature is flooded with articles and books for designing RF filters applying the simulation techniques. Commercially, a variety of EM simulators are available from many sources such as Ansoft, CST, Semcad and Remcom. Finite element method, Finite difference method and Finite difference time domain are some of the techniques used by the simulators. Considering the number of variables for optimisation, appropriate technique should be selected for faster analysis and accurate design of filters [18–21]. The design variables are the mechanical dimensions of the printed patterns.

Solving problems with EM simulators is an involved process and in-depth understanding of electromagnetics is required to achieve the desired results [19]. Literature provided by the suppliers of EM simulators and general literature on EM microwave software support the understanding of electromagnetics for designing RF filters. Setting the variables for optimisation, constraints or the boundary limits for the variables and accessing the library of characterised standard components are explained in detail in the literature of suppliers. The simulators indicate all the required filter parameters for the fabrication of filters at the end of optimisation. EM simulators are also used for designing other passive components such as filters with dielectric resonators, multiplexers, couplers and antennas [19].

References

1. Landee RW, Davis DC, Albrecht AP (1957) Electronic designers handbook. McGraw-Hill, NY
2. Matthei GL, Young L, Jones EMT (1980) Microwave filters, impedance-matching networks, and coupling structures. Artech House, London
3. Borji A, Busuioc D, Safavi-Naeini S, Chaudhuri SK (2002) ANN and EM based models for fast and accurate modeling of excitation loops in combline-type filters, microwave symposium digest, 2105–2108
4. Arndt F, Brandt J (2002) MM/FE CAD and optimization of rectangular combline filters, microwave conference, pp 1–4
5. Yaot H-W, ZakiS KA, Atiat AE, Dolantt T (1997) Improvement of spurious performance of combline filters, microwave symposium digest, 1099–1103
6. Yu M (2007) Power handling capability for RF filters. IEEE Microwave Mag 8:88–97
7. Hoffman GR (1975) Resonated combline band pass filters, microwave conference, pp 54–58
8. Schoenlinner # B, Prechtel # U, Schumacher H (2010) Miniaturized ku–band filters in LTCC technology, evanescent mode versus. Standard cavity filters, Proceedings of the 40th european microwave conference, France pp 5–8

9. Leon K, Mazierska J, Jacob MV, Ledenyov D (2002) Comparing unloaded q-factor of a high-q dielectric resonator measured using the transmission mode and reflection mode methods involving s-parameter circle fitting IEEE MTT-S Digest, 1665–1668
10. Kemppinen E, Hulkko J, Leppavuori S (1987) Realization of integrated miniature ceramic filters for 900 MHz cellular mobile radio application, MW conference, 175–180
11. Mansour RR (2004) Filter technologies for wireless base stations. IEEE Microwave Mag 5:68–74
12. Mansour RR (2009) High-q tunable dielectric resonator filters. IEEE Microwave Mag 10:84–98
13. Wang C, Zaki KA (2007) Dielectric resonators and filters. IEEE Microwave Mag 8:115–127
14. Bahl IJ, Trivedi DK (1977) A Designer's guide to micro-stripline. Microwaves 16:174–182
15. Standard IPC-214A (2004) Controlled impedance circuit boards and high speed logic design, institute for interconnection and packaging electronics
16. Yu M, Panariello A, Ismail M, Zheng J (2008) 3-D EM simulators for passive devices. IEEE Microwave Mag 9:50–61
17. Ong CY, Weldon M, Quiring S, Maxwell L, Hughes MC, Whelan C, Okoniewski M (2010) Speed it up. IEEE Microwave Mag 11:70–78
18. Swanson D, Macchiarella G (2007) Microwave filter design by synthesis and optimisation. IEEE Microwave Mag 8:55–69
19. Shin S, Kanamalaru S (2007) Diplexer design using em and circuit simulation techniques. IEEE Microwave Mag 8:77–82
20. Ming Yu, Panariello Antonio, Ismail Mostafa, Zheng Jingliang (2008) 3-D EM simulators for passive devices. IEEE Microwave Mag 9:50–61
21. Williams Lawrence, Rousselle Steve (2008) EM at the core of complex microwave design. IEEE Microwave Mag 9:97–104

Chapter 5
Planning the Design of RF Filters

Abstract Designing RF filters that just satisfies customer specified requirements is termed as functional design of filters. Functional design is the core design task but it alone is not adequate in industrial environment. Filters should be designed for repeatability, reproducibility, producibility and reliability in addition to complying with functional design requirements. Such a design is termed as value added functional design. The benefits of value added design, the technical activities for realising value added functional design and the methods of integrating the four abilities with functional design are explained in detail with practical examples.

5.1 Need for Planning Design

The design and development of RF filters should ensure that the specified requirements (electrical, mechanical and environmental) of customer are satisfactorily met and compliance to the requirements is demonstrated to the customer. Such a design which just satisfies customer specified requirements is termed as functional design of filters. Functional design of a filter is definitely the core design task but it alone is not adequate. The inadequate functional design is shown in Fig. 5.1. RF filters should be designed for repeatability, reproducibility, producibility and reliability requirements in addition to complying with functional design requirements.

Simple explanations could be given for the four abilities.

1. Repeatability refers to verifying the compliance of electrical performance of filters for many (five or more) times by one operator.
2. Reproducibility refers to verifying the compliance of electrical performance of filters by many (five or more) operators.
3. Produceability refers to the ease of manufacturing the filters.

D. Natarajan, *A Practical Design of Lumped, Semi-Lumped and Microwave Cavity Filters*, Lecture Notes in Electrical Engineering,
DOI: 10.1007/978-3-642-32861-9_5, © Springer-Verlag Berlin Heidelberg 2013

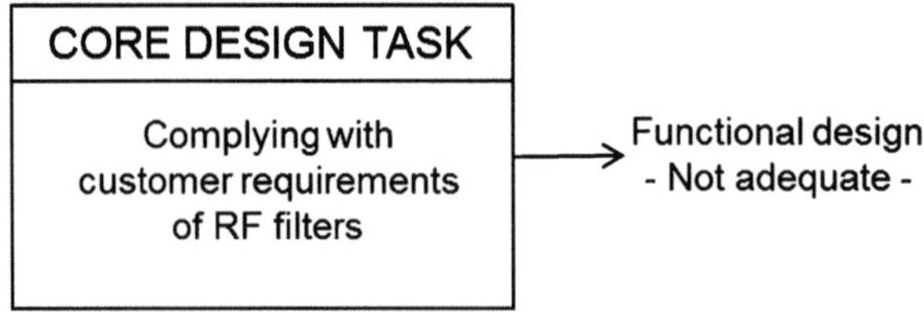

Fig. 5.1 Functional design of RF filter

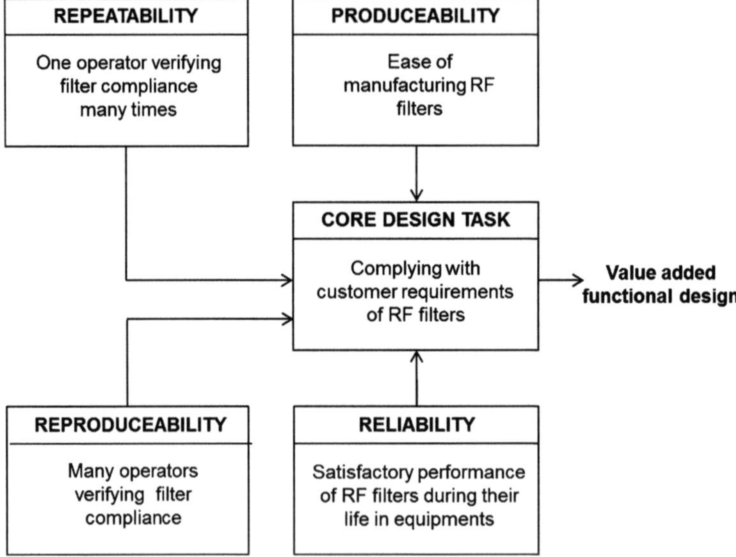

Fig. 5.2 Value added functional design of RF filter

4. Reliability refers to satisfactory performance during the life of equipments in
   which the filters are used.

The four abilities add values to functional design and the value added design
concept is shown in Fig. 5.2.

5.1.1 *Benefits of Value Added Functional Design*

The four abilities definitely add business values to the functional design of RF
filters and the value added are:

1. Repeatability and reproducibility ensure acceptance of filters during manufacturing and incoming inspection at customer premises, providing in-house
   and customer confidence.

Table 5.1 Suggested activities for the design of RF filters

Sl. No.	Design and development activities
1	Understanding customer requirements
2	Setting internal specifications
3	Computations of filter parameters
4	Design drawings and their review
5	Organising samples of RF filters
6	Internal testing of filters and validation by customer
7	Release of manufacturing drawings

2. Producibility enhances productivity and facilitates the ease of controlling quality during manufacturing. Manufacturing processes output filters with consistent electrical performance.
3. Reliability design ensures compliance to electrical performance during environmental tests and acceptable degradation of electrical performance of filters during the life of equipments in which the filters are mounted.

Integrating the four abilities with functional design should form the total objectives of RF filter design. The integration of the abilities with the design of RF filters is relatively easier and effective compared to complex electronic equipments. Planning the tasks during the design phase of filters is the key to the successful integration of the four abilities with the functional design of the filters.

5.2 Planning Design Phase

Planning is the process of listing what need to be done and how they have to be done for the realisation of an objective. Planning enables to achieve the objective with comfort. The technical and managerial tasks in planning are listed below and the technical task is explained in detail.

Technical task:

1. Listing the required design and development activities for achieving the desired objective

Managerial tasks:

2. Identifying the required managerial/material resources for the activities
3. Assigning a time schedule appropriate for the activities.

There could be variations in identifying the activities that would be required for achieving the objective of RF filter design. However, the minimum activities that are acceptable for the design phase of RF filters are shown in Table 5.1.

The activities listed above could be easily applied during the design phase of RF filters. For example, the activities could be planned and completed within two to 3 weeks for the microwave cavity band pass filter, shown in Table 3.3, assuming 50 % of the time is required for the fabrication of filter samples. The design and development activities are explained with their linkages to repeatability, reproducibility, producibility and reliability.

5.3 Understanding Customer Requirements

As mentioned earlier, customers indicate the electrical, mechanical and environmental interfacing requirements of RF filters in the form of specifications. The specifications form the basic design inputs. The specifications should be understood with clarity both by customers and filter designers. Ambiguities, if any, should be sorted out and the same should be recorded by filter designers even if they are sorted out verbally with customers.

For example, assume that customer has missed to specify the thickness of printed circuit board (PCB) on which filters are surface mounted. The thickness of PCB is obtained from customers so that the terminals for the surface mounting of the filters are positioned properly from the base of the filter. There could be inadvertent errors also in the specifications. Dwelling on the same example of surface mount filters, customer would have specified the dimensions of the tab terminals arbitrarily for surface mounting. It is difficult to connect the filters with tab terminations to Network analyser for tuning or for customer inspection. Hence, specially designed coaxial connectors are mounted temporarily on to the tab terminals for tuning and inspection purposes. Such connectors are called field replaceable coaxial connectors. To facilitate the mounting of the connectors, the dimensions of tab terminations should be aligned with the interface (mating) dimensions of the field replaceable coaxial connectors through discussions with customers. The use of field replaceable connectors improves the produceability of RF filters and eliminates unnecessary mechanical and thermal stresses to the terminals preventing the degradation of reliability of filters during customer inspection. Understanding customer requirements is really knowledge sharing. Customers would be delighted to provide all the required information and special test equipments, if any.

5.4 Setting Internal Specifications

5.4.1 Need for Internal Specifications

Design of RF filters is a process. RF filters that just satisfy the customer specified electrical performance requirements are considered as 3σ capability filter. RF filters with 3σ capability are bound to fail due to minor or inherent variations that would occur during manufacturing or at customer end. Variations caused by operators and test equipments are examples of inherent variations. Hence, there is a need to evolve internal electrical specification from customer specified requirements.

Table 5.2 Measured values of insertion loss	Sl. No.	Insertion loss (dB)
	1	0.97
	2	0.91
	3	0.91
	4	0.97
	5	0.94
	6	0.95
	7	0.93
	8	0.94
	9	0.94
	10	0.91

Setting internal electrical specifications better than the customer specified requirements for RF filters pushes up the capability of filters towards 6σ by design and it significantly enhances repeatability, reproducibility, producibility and reliability of RF filters. The quantum of improvements or margins that could be set for the internal electrical specification over the customer specified requirements primarily depends upon the capabilities of design team and the availability of design cycle time. Evolving internal specification is explained for the insertion loss characteristic of RF filters and it is applicable for all the characteristics.

5.4.2 Statistical Method of Setting Internal Specifications

Let the customer specification limit for the insertion loss of RF filter is 1 dB maximum. The measured values of insertion loss on ten filters are shown in Table 5.2. The measured values are within the specified limit of 1 dB maximum. For statistical analysis, 25 or more observations would be required but it is limited to ten observations for explaining the concept.

Customer specification limit: 1 dB maximum

$$\text{Mean } (\mu) = 0.937\,\text{dB}$$

$$\text{Standard deviation } (\sigma) = 0.022\,\text{dB}$$

Assuming that the observed insertion loss data is Normally distributed, the percentage of filters that are expected to fail in insertion loss could be estimated. The standardised variable [1], $(x-\mu)/\sigma$, is first calculated and then the area under the Standard Normal distribution is obtained from statistical Table. From the area the expected rejection could be estimated.

$$x = \text{Customer specification limit} = 1\,\text{dB maximum}$$

$$(x - \mu)/\sigma = (1 - 0.937)/0.022 = 2.86$$

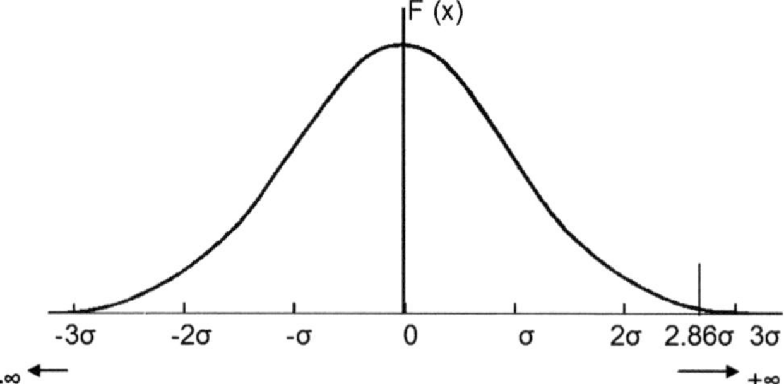

Fig. 5.3 Standard Normal distribution (Insertion loss:1 dB max)

Table 5.3 Measured values of insertion loss after improvement	Sl. No.	Insertion loss (dB)
	1	0.88
	2	0.90
	3	0.88
	4	0.85
	5	0.89
	6	0.85
	7	0.86
	8	0.90
	9	0.88
	10	0.85

Referring to statistical Table, approximately 0.21 % (the area to the right of 2.86σ) of the filters is expected to fail although the measured values are within the customer specified limit of 1 dB max. The Standard Normal distribution indicating the ordinate (2.86σ) for the rejection of 0.21 % is shown in Fig. 5.3. Additional information about the statistical analysis is available in Appendix-2.

It is obvious from Fig. 5.3 that additional design efforts are required for minimising the expected failures in insertion loss to an acceptable level. Assume 10 % margin over the customer specified limit of 1 dB maximum and the internal specification limit for insertion loss becomes 0.9 dB maximum. Let the measured values of insertion loss on ten filters be as in Table 5.3.

Customer specification limit: 1 dB maximum

Internal specification limit: 0.9 dB maximum

$$\text{Mean } (\mu) = 0.874\,\text{dB}$$

$$\text{Standard deviation } (\sigma) = 0.019\,\text{dB}$$

The measured values just satisfy the internal specification limit of 0.9 dB maximum and they are well within the customer specification limit of 1 dB maximum. The expected rejection is estimated against the customer specification limit.

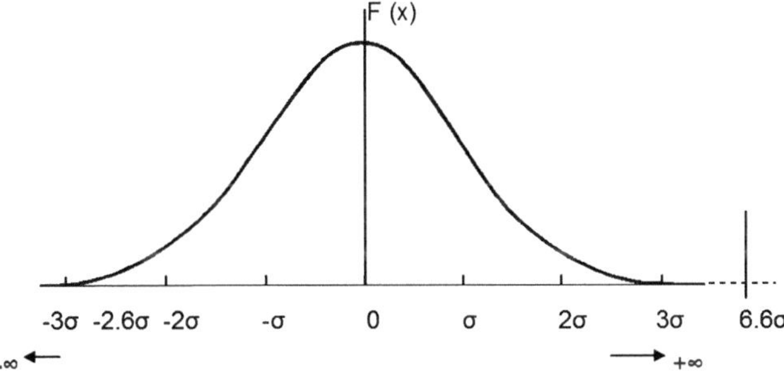

Fig. 5.4 Standard Normal distribution (Insertion loss: 0.9 dB max)

x = Customer specification limit = 1 dB maximum

$$(x - \mu)/\sigma = (1 - 0.874)/0.019 = 6.6$$

Considering inherent variations, only three filters per million filters (≈ 3 ppm) are expected to fail and it is acceptable. The Standard Normal distribution indicating the ordinate (6.6σ) is shown in Fig. 5.4. The example confirms that there is a need to evolve internal electrical specifications from customer specified requirements.

5.4.3 Recommended Internal Specification Limits

A minimum of 10 % margin is adequate for insertion loss and return loss characteristics of RF filters. The required margins in the internal specifications of insertion loss and return loss characteristics are achieved during the tuning of filters and they are not used in the computation of filter parameters.

For pass band, the design margin is decided based on the pass band edge frequencies. For example, if the customer specified pass band is (10–30) MHz for low pass filter, the internal specification could be fixed as (10–32) MHz. However, 30 MHz is used for calculating design parameters and the required margin is realised during the tuning of the filter. For rejection characteristic, 10 to 15 dB margins is recommended for the purpose of calculating design parameters but 5 to 10 % margin would be adequate during the tuning of filters. In the design of RF filters, it is easier to control the standard deviation of electrical characteristics than achieving higher level of margin for the characteristics. An example of band pass filter is shown in Table 5.4 with customer and internal specification limits.

Setting internal margins need not be limited to electrical characteristics only. For example, if the depth of blind threaded holes for mounting is specified as 4 mm to customers, it could be specified as 5 mm minimum in internal machine drawings. Many customers do not specify vibration test. However, transportation vibration test is included in the internal environmental specifications to ensure that the tuning of

Table 5.4 Specification of a band pass filter with customer and internal limits

Parameter	Customer limit	Internal limit
Pass band centre frequency, fo	1 GHz	1 GHz
Bandwidth	80 MHz	85 MHz
Insertion loss	2 dB max	1.8 dB max
Return loss	20 dB min	22 dB min
Rejection at (fo $\pm$ 100) MHz	60 dB min	65 dB min
Pass band ripple	0.4 dB max	0.2 dB max
Operating temperature	0–50°C	0–50°C
Input/Output connectors	SMA (f)	SMA (f)

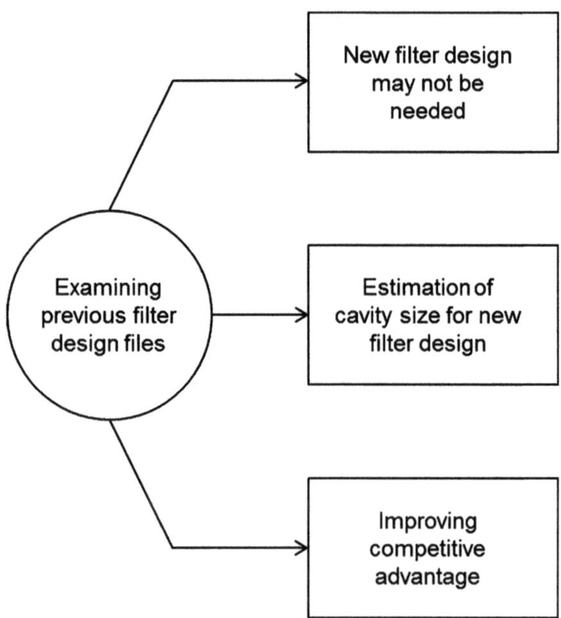

Fig. 5.5 Benefits of examining previous filter design files

filters is not disturbed during the transportation of filters to customers. Telecom or military environmental testing standard could be referred for obtaining the test conditions for the vibration test. The finalised internal specifications for RF filters are documented and the specifications form the *real* inputs for the design of the filters.

5.5 Computations of Filter Parameters

For lumped/semi-lumped filters, the filter parameters are the values of inductors and capacitors. For microwave cavity filters, the design parameters are the mechanical dimensions of cavity block, resonators and other relevant dimensions. The computation of filter parameters is explained in Chaps. 6 and 7. Before proceeding with the computation of filter parameters, previous design files on RF filters with similar performance characteristics are examined and relevant design information is extracted for possible application to the new design of filters. Information in the previous design files could be computerised for the fast retrieval of relevant infor- mation. Examining the previous design files has excellent benefits and the benefits are explained. The summary of the benefits is presented in Fig. 5.5.

1. Redesign efforts might not be required for the new filter. For example, the performance requirements of the new filter could be met by minor changes in previously designed filters or by re-tuning .
2. If new design efforts are not needed, the available design time could be used to apply new design ideas and techniques on the previously designed filters to improve margins on electrical performance, thus enhancing competitive advantage.
3. By examining the insertion loss/bandwidth of previously designed microwave cavity filters, approximate size of microwave cavity could be decided for new filters

5.6 Design Drawings and Their Review

Design drawings are prepared using the computed filter parameters. A minimum of two categories of drawings are prepared, one for the piece parts of filter and the other for the assembly of filter using the piece parts. The drawings of mechanical piece parts represent the design of the piece parts of a filter for machining. The design drawings are reviewed for improving the produceability and reliability of RF filters.

5.6.1 Sources of Data for Review

Innovation and creativity are applied in the design of the piece parts to achieve the producibility and reliability of filters by design. Manufacturing data on similar RF filters are analysed to support the creative efforts in the preparation of the design drawings. The sources of the data on similar RF filters and the benefits are shown in Fig. 5.6 and the same are explained with examples.

5.6.1.1 Failure Data on Similar RF Filters

The causes of failures observed during manufacturing and those reported by cus- tomers on the similar RF filters should be eliminated in the design of new filters, preventing the occurrence of the failures in the new filters. For example, if the manufacturing records of a filter indicate that unacceptable variation in insertion loss was caused by poor RF contact between the resonators and the cavity block of the

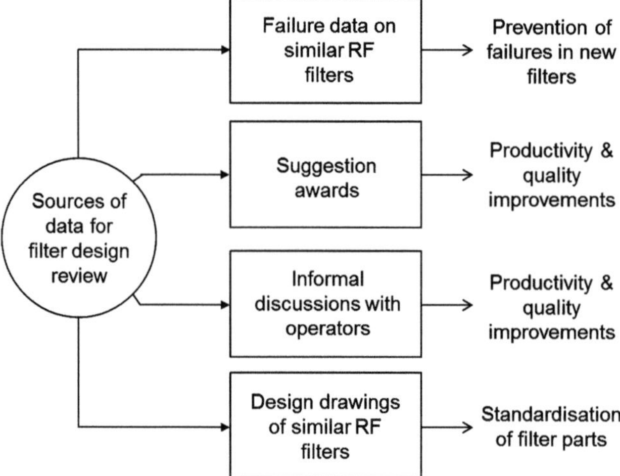

Fig. 5.6 Sources of data for review and their benefits

filter, the design of mounting of resonators could be improved or surface finish for the mounting surfaces could be specified in piece part drawings for new filters.

5.6.1.2 Suggestion Awards

It is quite common in industries to have suggestion awards for operators. Awards are given for productivity and quality improvements. Reports on suggestion awards related to RF filters are examined for incorporating the suggestions in the design of new filters.

5.6.1.3 Informal Discussions

Informal discussions are held with the operators in manufacturing areas. More information is obtained through informal discussions rather than involving them in formal design reviews. For example, one operator might inform that flux residue remains locked in some components of L-C filter PCB, expressing difficulty in cleaning. The useful information from the operator is that it is not easy to produce filters and the filters with flux residue might be despatched to customers affecting the reliability of filters during its useful life. Utilising the information from the operators, the feasibility of improving the layout of reactive components in new filters could be explored to eliminate the observed difficulty in post-cleaning operation of assembled PCB.

5.6.1.4 Manufacturing Drawings

Manufacturing drawings of similar RF filters are examined with the view to ensure standardisation of piece parts in new filters. Standardisation is controlling the proliferation of the piece parts. For example, design efforts are made to use the piece parts of previously designed RF filters in new filters. The piece parts that are examined are printed circuit boards, capacitors, metallic enclosures, resonators and tuning/mounting fasteners. Standardisation of the piece parts of filters minimises unnecessary load on purchasing and storage processes, directly contributing for the produceability and quality of piece parts.

5.6.1.5 Formal Review of Design Drawings

After incorporating the relevant information from the manufacturing data of similar RF filters for the new filters, the design drawings of the filters are reviewed formally, involving the Heads of various departments in manufacturing. The suggestions of design review are recorded. They are analysed and appropriate decisions are taken for incorporating them in the new design of filters.

5.7 Internal Testing and Validation

Samples of RF filters are organised for internal testing and validation by customer. Parts are fabricated or procured as per design drawings. The parts are then assembled using filter assembly drawings. The filters are tuned for testing. RF filter samples are tuned and tested as per the internal specifications of the filters. The internal testing is also known as design verification testing. Test reports, if submitted to customers, should indicate the requirements specified by customers. Validation of RF filters is done by customers. Validation is verifying compliance to customer requirements and the functional use of the filters by mounting them in equipments.

The failures observed during internal verification testing and those reported by customers during validation testing are recorded. After cause analysis, suitable corrective actions are implemented in the design drawings of filter with revision control. If required, verification and validation testing are repeated to confirm the efficacy of the corrective actions. After the satisfactory completion of internal testing and customer validation testing, the drawings are released for the manufacturing of filters. The set of drawings of RF filter is the design output. Incidentally, the design file containing the records for the activities listed in Table 5.1 in the design phase of RF filters satisfies the design and development requirements of ISO9001 Quality Management System.

Reference

1. Chatfield C (1983) Statistics for technology, a course for applied statistics. Chapman and Hall, London

Chapter 6
Design of Lumped and Semi-Lumped RF Filters

Abstract The Microwave Filter Design Book Matthei (Microwave filters, impedance-matching networks, and coupling structures. Artech House, 1980) provides established design equations for the design of Tchebyscheff lumped and semi-lumped tubular low pass filters. The design equations are organised in the form of tutorials and illustrated with examples. The successful industrial experience in complying with customer requirements in the frequency band, (20–70) MHz for lumped and (150–250) MHz for semi-lumped tubular low pass filters are blended with the illustrations. Software could be developed for the tutorials. Guidance for the selection of filter parts, layout, end coupling and tuning of filters is provided.

6.1 Introduction

The design of lumped and semi-lumped RF filters is the process of converting the electrical specifications of the filters into the values of inductors and capacitors using design equations. The Microwave Filter Design Book [1] provides established design equations for computing the values of reactive elements both for maximally flat and Tchebyscheff types of filters. The application of the design equations is successful in complying with customer requirements in the frequency band, (20–70) MHz for lumped and (150–250) MHz for semi-lumped tubular low pass filters. The extension of the frequency bands is also feasible using the design equations of the Microwave Filter Design Book. The design equations are organized in the form of tutorials and the design of Tchebyscheff lumped and semi-lumped tubular low pass filters are illustrated with examples. Software or Excel tool could be developed for the tutorials to save design time. Selection of the parts of filters is equally important to ensure satisfactory environmental and reliable performance of the filters. Guidance for the selection of parts is presented before explaining the computational procedures.

D. Natarajan, *A Practical Design of Lumped, Semi-Lumped and Microwave Cavity Filters*, Lecture Notes in Electrical Engineering,
DOI: 10.1007/978-3-642-32861-9_6, © Springer-Verlag Berlin Heidelberg 2013

6.2 Lumped RF Filters

6.2.1 Selection of Filter Parts

Inductors and capacitors are the most important parts in lumped filters. The other parts are printed circuit board, input/output coaxial connectors, box and fasteners. The Q-factors of inductors and capacitors are critical to satisfy the insertion loss of filters. Single layer air-cored coils are adequate as the values of inductors are quite low at 20 MHz.

Designing coils by using silver plated (3 to 5 μ) copper wire with higher diameter and lower length ensures high Q-factor for the coils. Increasing the internal diameter and reducing the pitch of coils minimises the length of copper wire required for fabrication. Enamelled copper wire or chip inductor is used as an alternative to silver plated copper wire if the specified performance of filters could be achieved. Hand or machine wound air-cored inductors are preferred as they facilitate the adjustment of the pitch of inductors for the tuning of filters. Inductors are directly soldered on to printed circuit boards. If inductors require a mechanical support, PTFE or ceramic rods are used as support mandrel as they do not reduce the Q-factor of the inductors significantly.

Capacitors with many dielectric materials such as ceramic, polyester, polypropylene, mica and PTFE are available both in leaded and chip forms. Multi-layer ceramic chip capacitors are widely used in filters as they are characterised by small size and high performance. The characteristics that are considered in the selection of capacitors are:

1. Dielectric loss factor (tan δ)
2. Load life stability
3. Temperature coefficient
4. Operating temperature range.

Dielectric loss factor (tan δ) of capacitors directly influences the insertion loss of filters. Capacitors with low values of tan δ are selected. Load life stability indicates the change of capacitance with usage life period. Capacitors with better load life stability provide stable performance of filters in equipments. Ceramic capacitors are available with negative, zero and positive temperature coefficients. Temperature coefficient indicates the rate of change of capacitance with temperature. Ceramic capacitors with negative or zero temperature coefficients are suited to ensure the stability of performance over the operating temperature range of filters. Capacitors are rated for wider operating temperature range and hence no special consideration is required in the selection of capacitors. The catalogues and application notes of leading chip capacitors are quite useful for understanding the characteristics of capacitors.

Glass epoxy copper clad sheet, Grade FR4, is adequate for mounting the inductors and capacitors of lumped RF filters. Input and output coaxial connectors are usually BNC or TNC or N connector series. The connectors with extended

Table 6.1 Sample requirements of lumped low pass filter with random values

Filter characteristics	Customer limit	Internal limit
Pass band	(10–30) MHz	(10–32) MHz[a]
Pass band insertion loss	1 dB max.	0.9 dB max.
Return loss	20 dB min.	22 dB min.
Rejection at 40 MHz	50 dB min.	55 dB min.[b]
Input/output impedance	50 Ω	50 Ω
Operating temperature	0–50 °C	0–50 °C
Input/output connectors	SMA (f)	SMA (f)

For calculating the low pass filter parameters:
[a] Recommended 30 MHz as pass band edge frequency for design calculations. Limit of 32 MHz to be ensured while tuning the filter
[b] Recommended 65 dB as rejection at 40 MHz for design calculations. 55 dB to be ensured while tuning the filter

dielectric equal to the wall thickness of metal box are used to avoid discontinuities in input/output impedance. Feed thorough terminals are used for the surface mounting of filters. Appendix-3 could be referred for additional information about coaxial connectors with extended dielectric and feed through terminals.

6.3 Illustrated Design of Lumped RF Filter

The design of Tchebyscheff lumped RF low pass filter is illustrated with the sample electrical specification, shown in Table 6.1. Both customer and internal specification limits are shown.

6.3.1 Computational Steps for Filter Design Parameters

Filter design involves many intermediate computations for determining the values of inductors and capacitors. The steps in the computations are:

1. Determining the number of filter elements

   (a) Calculate ω' / ω_1', the ratio of rejection frequency to the upper pass band edge frequency in radians
   (b) Calculate ε
   $\varepsilon = [\text{antilog}_{10}\ (L_{Ar}/10)] - 1$, where L_{Ar} is the ripple in pass band
   (c) Calculate the number of filter elements, n

2. Computing the values of filter elements

   (d) Compute g-values (element values of low-pass proto-type filters)
   (e) Compute the normalised values of filter elements
   (f) Compute the actual values of filter elements

6.3.2 Number of Filter Elements

Referring to the filter specifications shown in Table 6.1:

f_u, Pass band upper edge frequency $= 30$ MHz
f, rejection frequency $= 40$ MHz

6.3.2.1 Calculate ω'/ω_1'

$$\omega'/\omega_1' = f / f_u = 40/30 = 1.33$$

6.3.2.2 Calculate ε

Assume L_{Ar}, pass band ripple of 0.2 dB for design calculations.
$\quad \varepsilon = [\text{antilog}_{10} (L_{Ar}/10)] - 1 = [\text{antilog}_{10} (0.2/10)] - 1 = 0.047$

6.3.2.3 Calculate the Number of Filter Elements, n

L_A, rejection (dB) at 40 MHz $= 65$

$n = \cosh^{-1}[\text{SQRT} \{(10^\wedge (L_A/10)-1/\varepsilon\}]/\cosh^{-1}(\omega'/\omega_1')$
$n = \cosh^{-1}[\text{SQRT} \{(10^\wedge (65/10)-1/0.047\}]/\cosh^{-1}(1.33) = 12.5$
n is rounded up to the next higher integer.
The number of filter elements, $n = 13$

6.3.3 Values of Filter Elements

6.3.3.1 Compute g-Values

The number of the g-values is related to the number of elements. For n elements, g_0 to g_{n+1} values are computed. For $n = 13$, g_0 to g_{14} values are computed. The steps and formulas for calculating the g-values are explained in Chap. 7, Design of microwave cavity filters. The calculated g-values are shown in Table 6.2. The g-values are slightly modified to ensure symmetry of the values about g_7 and the modified g-values are also shown in Table 6.2.

Table 6.2 g-Values for 13 sections

g_n	Calculated g-values for 0.2 dB ripple	Modified g-values for 0.2 dB ripple
g_0	1.0000	1.0000
g_1	1.3977	1.3977
g_2	1.4057	1.4057
g_3	2.3328	2.3328
g_4	1.5529	1.5529
g_5	2.4144	2.4144
g_6	1.5755	1.5755
g_7	2.4280	2.4280
g_8	1.5755	1.5755
g_9	2.4144	2.4144
G_{10}	1.5528	1.5529
G_{11}	2.3324	2.3328
G_{12}	1.4046	1.4057
G_{13}	1.3889	1.3977
G_{14}	1.0000	1.0000

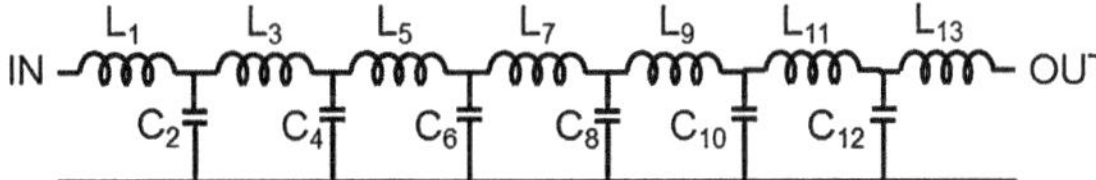

Fig. 6.1 Low pass filters (13 elements)

6.3.3.2 Compute the Normalized Values of Filter Elements

The 13 elements of T-section low pass filter with filter input and filter output are shown in Fig. 6.1, comprising of seven inductors and six capacitors.

The g-values, g_1 to g_n, represent the values of the filter elements in Henries (H) and Farads (F) of a normalised low pass filter with 1 Hz as pass band upper edge frequency ($f'_u = 1\text{Hz}$) and such a filter is defined as prototype low pass filter [1]. The g-values, g_0 and g_{n+1}, represent generator and load resistance/conductance as applicable or in other words, they represent input/output impedance of the prototype low pass filters. The values of the 13 elements and the input/output impedance for the prototype low pass filters are shown in Table 6.3 with temporary identifications, L'_1, C'_2, L'_3 etc.

6.3.3.3 Compute the Actual Values of Filter Elements

The element values of the prototype filter are scaled up or converted to the specified higher input/output impedance, 50 Ω, and pass band upper edge frequency, 30 MHz (f_u).

Table 6.3 Normalised values of low pass filter elements

g_n	g-Values for 0.2 dB ripple	Element values of prototype low pass filter
g_0	1.0000	Input impedance, $R_0' = 1\ \Omega$
g_1	1.3977	$L_1' = 1.3977$ H
g_2	1.4057	$C_2' = 1.4057$ F
g_3	2.3328	$L_3' = 2.3328$ H
g_4	1.5529	$C_4' = 1.5529$ F
g_5	2.4144	$L_5' = 2.4144$ H
g_6	1.5755	$C_6' = 1.5755$ F
g_7	2.4280	$L_7' = 2.4280$ H
g_8	1.5755	$C_8' = 1.5755$ F
g_9	2.4144	$L_9' = 2.4144$ H
g_{10}	1.5529	$C_{10}' = 1.5529$ F
g_{11}	2.3328	$L_{11}' = 2.3328$ H
g_{12}	1.4057	$C_{12}' = 1.4057$ F
g_{13}	1.3977	$L_{13}' = 1.3977$ H
g_{14}	1.0000	Output impedance, $R_{14}' = 1\Omega$

The expressions for the conversions:

Input impedance, $R_0 = 50^*R_0' = 50^*1 = 50\ \Omega$

Output impedance, $R_{14} = 50^*R_{14}' = 50^*1 = 50\ \Omega$

$\omega_1'/\omega_1 = 1/(2^* \prod{}^* f_u) = 1/(2^* \prod{}^* 30E10^6)$

$L = (R_0/R_0')(\omega_1'/\omega_1)^*L'$

$C = (R_0'/R_0)(\omega_1'/\omega_1)^*C'$

$L1 = (50/1)[1/(2^* \varPi^*30E10^6)]^*1.3977^*10^6 = 0.37\mu H$

$C2 = (1/50)[1/(2^* \varPi^*30E10^6)]^*1.4057^*10^{12} = 149.2 pF$

The converted values are the element values of low pass filter defined in Table 6.1 and the values are tabulated in Table 6.4.

The internal diameter and the pitch of single layer air-cored coils are calculated for the inductance values using the expression in Sect. 6.4 at the end of the chapter. Multilayer ceramic chip capacitors with zero temperature coefficient (NPO type) are suitable for the low pass filter. The nearest standard values are chosen for the capacitors. 150 pF is used for 149.2 pF and two standard capacitors, 150 and 18 pF, are paralleled for all other capacitor values. The low pass filters circuit with inductance values of coils and standardised values of capacitors is shown in Fig. 6.2.

6.3.4 Layout Design for Filter Elements

Layout design of filter elements is the method of mounting the elements on a PCB. The layout design should:

Table 6.4 Element values of low pass filter

Element values of prototype low pass filter	Element values of low pass filter, defined in Table 6.1
Input impedance, $R_0' = 1\ \Omega$	Input impedance, $R_o = 50\ \Omega$
$L_1' = 1.3977$ H	$L_1 = 0.37\ \mu$H
$C_2' = 1.4057$ F	$C_2 = 149.2$ pF
$L_3' = 2.3328$ H	$L_3 = 0.62\ \mu$H
$C_4' = 1.5529$ F	$C_4 = 164.9$ pF
$L_5' = 2.4144$ H	$L_5 = 0.65\ \mu$H
$C_6' = 1.5755$ F	$C_6 = 167.3$ pF
$L_7' = 2.4280$ H	$L_7 = 0.64\ \mu$H
$C_8' = 1.5755$ F	$C_8 = 167.3$ pF
$L_9' = 2.4144$ H	$L_9 = 0.65\ \mu$H
$C_{10}' = 1.5529$ F	$C_{10} = 164.9$ pF
$L_{11}' = 2.3328$ H	$L_{11} = 0.62\ \mu$H
$C_{12}' = 1.4057$ F	$C_{12} = 149.2$ pF
$L_{13}' = 1.3977$ H	$L_{13} = 0.37\ \mu$H
Output impedance, $R_{14}' = 1\ \Omega$	Output impedance, $R_{14} = 50\ \Omega$

Fig. 6.2 Low pass filters with element values

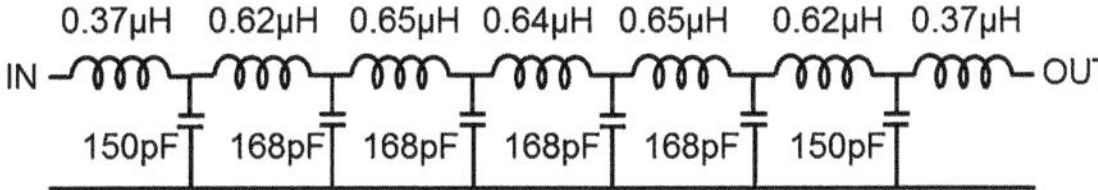

1. Minimise interconnection losses by eliminating unnecessary leads in the interconnection of elements
2. Minimise parasitic coupling between adjacent coils
3. Facilitate the tuning of filter by the tapping of coils.

The design for the layout of elements is relatively simpler for filters with operating frequency up to 100 MHz. For the low pass filter shown in Fig. 6.2, all the components are surface mounted and soldered on a 1.6 mm thick glass epoxy PCB. Microstrip transmission lines are designed for the mounting the capacitors and coils of high frequency filters. After soldering the components, the flux residues are cleaned using suitable PCB cleaning solvents such as isopropyl alcohol or trichloroethane before the residues harden. Flux residues increase the insertion loss of filters over a period of time. Precautions for cleaning flux residues of chip capacitors are explained in Chap. 8.

6.3.5 End Coupling

End coupling refers to the connecting of first and last elements of filter to the centre contacts of the input and output coaxial connectors of filter. The connectors are mounted on the sides of boxes such that their centre contacts are directly surface mounted to PCBs for soldering. Connectors are mechanically mounted before soldering to avoid mechanical stresses on the soldered joints of the connectors. Finally, the filter assembly is mounted in a suitable metal box with adequate height to minimise parasitic coupling of coils with the box. Boxes are silver plated or finished with conductive chromate conversion coating and painted.

6.3.6 Tuning of Lumped RF Filter

Lumped RF filter, when assembled, does not meet the specified electrical specifications. The values of inductors and capacitors need to be adjusted for satisfying the specifications. The process of adjusting the values of inductors and capacitors of a filter is termed as tuning of filters. Vector Network Analyser (VNA) is preferred for the tuning of filters. General information for tuning and about the VNA set-up is presented in Chap. 8.

The inductance values of coils are adjusted by varying the pitch of the coils. The interconnection tapping to the inductors are also changed for adjustment. Capacitance values are varied by paralleling suitable values capacitors. Initial tuning focuses for realising the pass band and approximate rejection specification of the filter. Tuning is started with input and output filter sections and it is continued with middle filter sections. Tuning is an iterative process and previously tuned sections require retuning due to adjacent or parasitic coupling effects. After the realisation of pass band, the filter is fine tuned for satisfying the return loss and rejection requirements, maintaining the pass band. Improving return loss reduces insertion loss. is continued to ensure that the internal specification of filters is satisfied with the cover plate assembled to box. After the tuning of filters, Silicone RTV (Room Temperature Vulcanisation) adhesive is applied over the coils to ensure the dimensional stability of the coils during handling, storage and transportation. The dielectric constant of the adhesive is close to 1.0 and it does not affect the electrical performance of filters.

6.4 Illustrated Design of Tubular Filter

A semi-lumped RF filter has a reactive network like lumped RF filters. Inductors are used in discrete form and capacitors are developed in distributed form. The design of Tchebyscheff semi-lumped tubular RF low pass filter is illustrated with the sample electrical specification shown in Table 6.5. Both customer and internal specification limits are shown.

Table 6.5 Requirements of tubular low pass filter with random limits

Filter characteristics	Customer limit	Internal limit
Pass band	(10–450) MHz	(10–465) MHz[a]
Pass band insertion loss	2.0 dB max.	1.8 dB max.
Return loss	18 dB min.	20 dB min.
Rejection at 600 MHz	30 dB min.	35 dB min.[b]
Input/output impedance	50 Ω	50 Ω
Operating temperature	0–50 °C	0–50 °C
Input/output connectors	N (f)	N (f)

For calculating the tubular low pass filter parameters:
[a] Recommended 450 MHz as pass band edge frequency for design calculations. Limit of 465 MHz to be ensured while tuning the filter
[b] Recommended 40 dB as rejection at 600 MHz for design calculations. Limit of 35 dB min. to be ensured while tuning the filter

Fig. 6.3 Tubular low pass filter (9 elements)

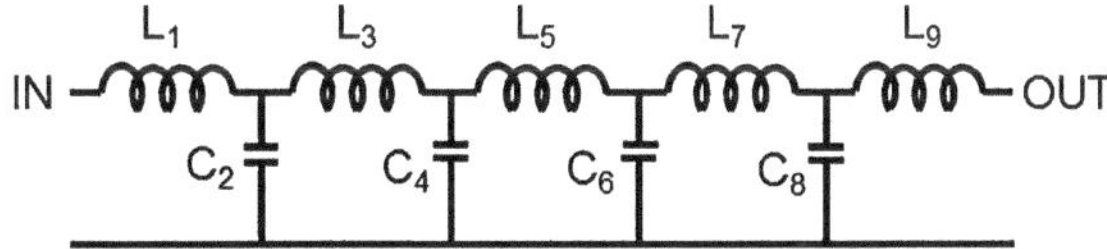

6.4.1 Number of Filter Elements

The design calculations of T-section tubular RF low pass filter are exactly same as that of lumped RF low pass filters. The required data for determining the number filter elements is reproduced from the filter specifications shown in Table 6.5.

f_u = 450 MHz
f, rejection frequency = 600 MHz

$$\omega'/\omega_1' = f/f_u = 600/450 = 1.33$$

Assume L_{Ar}, pass band ripple of 0.2 dB for design.
$\varepsilon = [\text{antilog}_{10}\,(L_{Ar}/10)] - 1 = [\text{antilog}_{10}\,(0.2/10)] - 1 = 0.047$
L_A, rejection (dB) at 600 MHz = 40
$n = \cosh^{-1}[\text{SQRT}\,\{(10^\wedge\,(L_A/10)-1/\varepsilon\}]/\cosh^{-1}(\omega'/\omega_1')$
$n = \cosh^{-1}[\text{SQRT}\,\{(10^\wedge\,(40/10)-1/0.047\}]/\cosh^{-1}(1.33) = 8.6$
n is rounded up to the next higher integer.
The number of filter elements, n = 9

The 9-element filter circuit with input and output is shown in Fig. 6.3.

Table 6.6 Element values of tubular low pass filter

g_n	g-Values for 0.2 dB ripple	Element values of prototype low pass filter	Element values of tubular low pass filter, defined in Table 6.5
g_0	1.0000	Input impedance, $R_0' = 1\ \Omega$	Input impedance, $R_o = 50\ \Omega$
g_1	1.3866	$L_1' = 1.3866$ H	$L_1 = 24.5$ nH
g_2	1.3937	$C_2' = 1.3937$ F	$C_2 = 9.9$ pF
g_3	2.3009	$L_3' = 2.3009$ H	$L_3 = 40.7$ nH
g_4	1.5337	$C_4' = 1.5337$ F	$C_4 = 10.8$ pF
g_5	2.3733	$L_5' = 2.3733$ H	$L_5 = 42.0$ nH
g_6	1.5337	$C_6' = 1.5337$ F	$C_6 = 10.8$ pF
g_7	2.3009	$L_7' = 2.3009$ H	$L_7 = 40.7$ nH
g_8	1.3937	$C_8' = 1.3937$ F	$C_8 = 9.9$ pF
g_9	1.3866	$L_9' = 1.3866$ H	$L_9 = 24.5$ nH
g_{10}	1.0000	Output impedance, $R_{10}' = 1\ \Omega$	Output impedance, $R_{10} = 50\ \Omega$

Fig. 6.4 Tubular low pass filter with element values

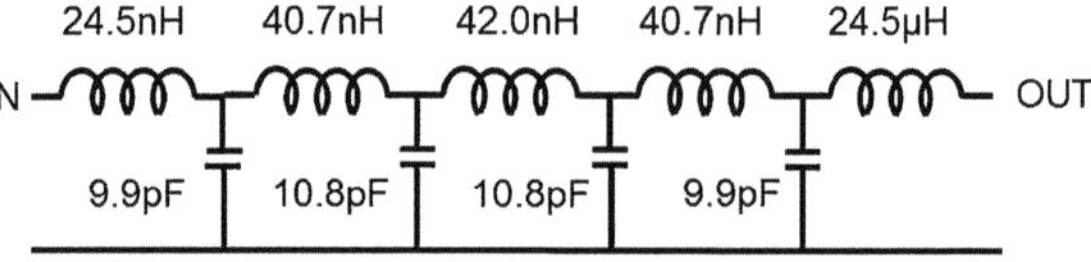

6.4.2 Values of Tubular Filter Elements

The g-values of the prototype low pass filter, element values of the prototype low pass filter and the actual element values of the tubular low pass filter are shown in Table 6.6 and in Fig. 6.4. The realisation of the elements is explained for the tubular low pass filter.

6.4.3 Realisation of Tubular Filter

A 50 Ω coaxial air line is required for the realisation of tubular low pass filter. N-jack coaxial connectors are preferred at both the ends of the line. The length and the diameter of the coaxial line are decided by the number of filter sections. In tubular filter, the centre conductor of the 50 Ω coaxial air line is not a single length of metallic conductor. The centre conductor consists of short lengths of metallic conductors that are interconnected by short lengths of PTFE insulators. For the tubular filter represented in Fig. 6.4, four metallic centre conductors and five PTFE insulators are interconnected to form the centre conductor of the coaxial air line.

The internal diameter of the outer conductor of 50 Ω coaxial air line is assumed to be 10 mm (D). The diameter of the centre contact (d) of the coaxial air line is calculated using the expression,

$$\text{Characteristic impedance, } Z_0 = \frac{60\ln(D/d)}{\sqrt{\varepsilon_r}} \text{ ohms}$$

$$d = 4.3\,\text{mm}$$

The five coils of the tubular filter are designed as explained in the Sect. 6.4 at the end of the chapter and the coils are inserted over the $\Phi 4.3$ mm PTFE centre conductor. The length of the insulators is arbitrarily fixed as 1.5 times the length of the wound inductors to facilitate varying the pitch of coils. The four capacitors are realised in distributed form between the four metallic centre conductors and the outer conductor of the coaxial air line.

The distributed capacitance developed between the metallic centre conductor and the outer conductor of the coaxial air line is calculated from the expression,

Capacitance [2] = 0.02414* ε_r/log (D/d) pF/mm
C_d = [0.02414 * ε_r/log (D/d)] * 1 pF
ε_r = Relative dielectric constant
D, Inner diameter of coaxial line = 10 mm
d, diameter of metallic centre conductor = 4.3 mm
Assume the length of the metallic centre contact, l = 5.5 mm
C_d = [0.02414 *1/log (10/4.3)] * 5.5 = 0.36 pF

The distributed capacitance, 0.36 pF, is inadequate for the tubular filter, represented in Fig. 6.4. Hence, a different mechanical design is required for realising the distributed capacitance but without disturbing the coaxial structure. The mechanical design is explained for one inductor-capacitor section [$L_1 - C_2$ section] of the tubular low pass filter and it is applicable for the remaining sections.

The value of the distributed capacitance, C_2, is 9.9 pF. The diameter of the metallic centre conductor (d) is increased, reducing the gap (t) between the centre and outer conductors of coaxial line. The gap is also filled with a dielectric material that has high dielectric constant (ε_r) and low loss (tanδ). The increased diameter of the metallic centre conductor is decided by the characteristics of dielectric material. Ultra low loss PTFE flexible laminate, 0.254 mm thick, having relative dielectric constant (ε_r) of 2.2 is a standard dielectric material. A clearance of 0.1 mm is required for inserting the filler material between the centre and outer conductors of the coaxial line.

t, thickness of PTFE flexible laminate = 0.254 mm
Dia. of metallic centre conductor = 10-[2*(0.254 + 0.1)] = 9.3 mm
Assume the length of the metallic centre conductor, l = 5.5 mm

$$C_d = [0.02414 * 2.2/\log(10/9.3)] * 5.5 = 9.3 \text{ pF}$$

The distributed capacitance is 9.3 pF against the requirement of 9.9 pF for C_2. The value is designed to be lower to facilitate the tuning of tubular filter. Tuning

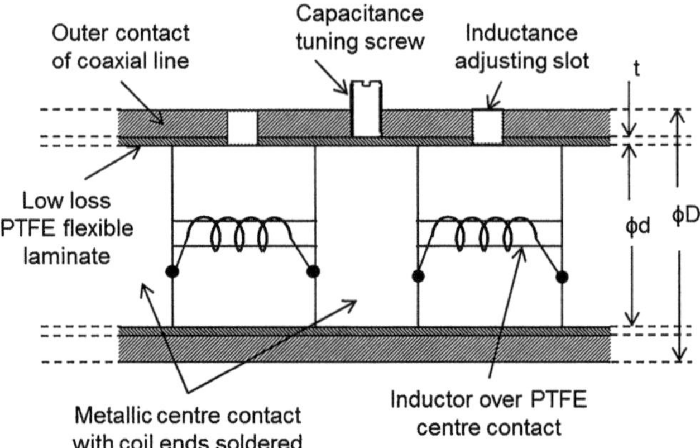

Fig. 6.5 Sections of tubular low pass filter

screws, $\Phi 4$ or $\Phi 5$ mm are fixed on the outer conductor of coaxial air line. The advancement of tuning screws slightly compresses the PTFE laminate increasing the distributed capacitance to the desired value. If required, the length of the metallic centre conductor could also be increased. A narrow slot is also provided on the outer conductor of the coaxial line above the coil to vary the pitch of the coil. The method of realising the design of the tubular filter is shown for one section in Fig. 6.5.

The terminals of the three middle coils and one terminal of the end coils are soldered to the adjacent metallic centre conductors. The other terminal of end coils is soldered to the centre conductors of input/output coaxial connectors. Innovation and creativity in the mechanical design of tubular filters significantly contribute for the produceability and reliability of the filters. The tubular filter is tuned after completing the assembly.

6.4.4 Tuning of Tubular Filter

The tubular filter is tuned by adjusting the inductance values of coils and distributed capacitors for satisfying the specifications of the filter. General information for tuning and about the VNA set-up is presented in Chap. 8. The values of inductors are varied by adjusting the pitch of the inductors through the narrow slots on the coaxial line. The distributed capacitors are adjusted by tuning screws. Care is exercised not to compress the PTFE flexible laminate excessively. The tuning procedure is same as that of the lumped RF filter, explained above. Tuning is started with input and output filter sections and it is continued with middle filter sections. Tuning is an iterative process and previously tuned sections require retuning due to coupling effects. Initial tuning focuses for realising the pass band and approximate rejection specification of the filter. After the realisation of pass

Fig. 6.6 Design dimensions
of coil

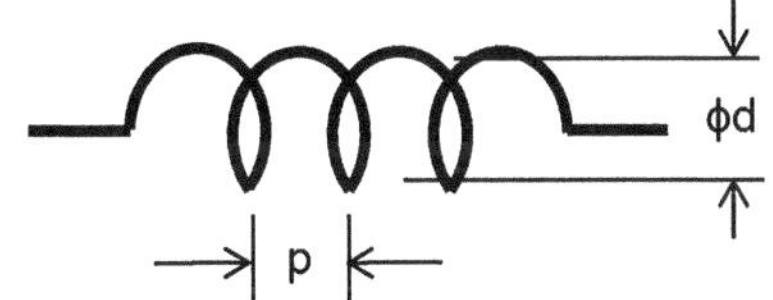

band, the filter is fine tuned for satisfying the return loss and rejection requirements, maintaining the pass band. Improving return loss reduces insertion loss also. After the tuning of filters, Silicone RTV (Room Temperature Vulcanisation) adhesive is applied over the coils to ensure the dimensional stability of the coils during handling, storage and transportation. The tubular filter is enclosed in another box, protecting the filter.

6.5 Design of Coils

Design of coils refers to determining the internal diameter (d), pitch (p) and the number of turns (n) shown in Fig. 6.6 to satisfy the inductance (L) of the coil.

Coils are designed using the expression [2] given below. The expression is accurate for the values of d/l up to 3 and it is adequate for most of the requirements of lumped and tubular filters.

$$n = 1016^{*}L^{*}p/d^2$$

n   Number of turns
L   Inductance in μH
p   Pitch in mm
d   Internal diameter of coil in mm
l   Length of coil in mm = n*p

References

1. Matthei GL, Young L, Jones EMT (1980) Microwave filters, impedance-matching networks, and coupling structures, Artech House
2. Landee RW, Davis DC, Albrecht AP (1957) Electronic designers' handbook. McGraw-Hill, NY

Chapter 7
Design of Microwave Cavity Filters

Abstract The Microwave Filter Design Book [1] provides established design equations for the design of Tchebyscheff combline and iris-coupled band pass cavity filters. The design equations are organised in the form of tutorials and illustrated with examples. The successful industrial experience in complying with customer requirements in the frequency band, (1.0–4.5) GHz for cavity filters are blended with the illustrations. Polynomials are fitted to the relevant design graphs using Lagrange interpolation technique to eliminate the need to refer the graphs Software could be developed for the tutorials including the polynomials. Guidance for the selection of filter parts, methods of end coupling and tuning of filters in frequency and time domain are provided.

7.1 Introduction

Design of cavity filters is the process of converting the electrical specifications of the filters into the mechanical dimensions of cavity block, resonators and other parts using design equations. The Microwave Filter Design Book [1] provides established design equations and graphs for computing the mechanical dimensions both for maximally flat and Tchebyscheff types of filters. The application of the design equations is successful in complying with customer requirements in the frequency band, (1.0–4.5) GHz for the cavity filters. The extension of the frequency bands is also feasible using the design equations of the Microwave Filter Design Book. The design equations and graphs available in the various chapters of the Design Book are organised in the form of tutorials and the design of Tchebyscheff combline and iris-coupled band pass filters are illustrated with examples. Polynomials are fitted to the relevant design graphs using Lagrange interpolation technique to eliminate the need to refer the graphs. Software or Excel tool could be developed for the tutorials for repetitive designs. Practical information for the design of parts of the cavity filters, types of end coupling and tuning of filters are presented in detail.

D. Natarajan, *A Practical Design of Lumped, Semi-Lumped and Microwave Cavity Filters*, Lecture Notes in Electrical Engineering,
DOI: 10.1007/978-3-642-32861-9_7, © Springer-Verlag Berlin Heidelberg 2013

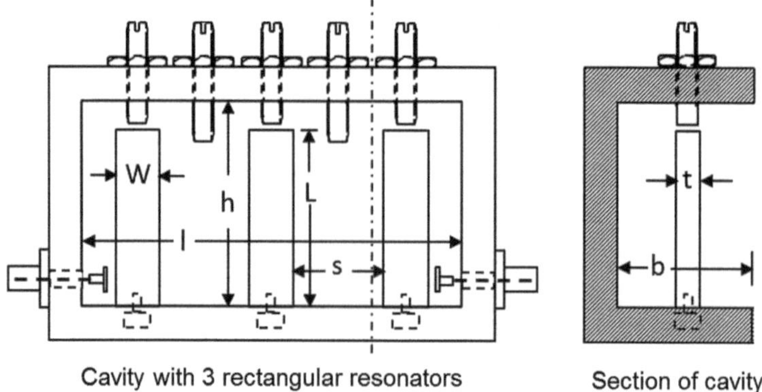

Fig. 7.1 Design dimensions of combline band pass filter

7.2 Combline Band Pass Filter

The general structure of combline band pass filter without side cover plate is shown in Fig. 7.1 indicating the mechanical dimensions that need to be computed for the piece parts of the filter. The number of filter sections (resonators) is limited to three in Fig. 7.1 for simplicity. Technical information for the design of the piece parts of combline filter is presented before computing the design parameters of filters.

7.2.1 Cavity Block

Cavity block is milled from a solid aluminium block with one side open. The inside corners of cavity block are finished with radius. Symbols, l, h and b designate the internal length, height and width of cavity. The wall thickness of cavity block is designed between 4 and 6 mm considering the availability of standard input/output coaxial connectors with extended dielectric.

7.2.2 Resonators

Bar or cylindrical resonators are used in cavity filters. Bar resonator has rectangular cross section with width, w, and thickness, t. Cylindrical resonator has circular cross section with diameter, d. Cylindrical resonators are preferred considering the produceability of cavity filters. One end of resonators is designed for mounting on cavity block and the other end is designed for developing distributed capacitance for tuning the resonators. Modified cylindrical resonators could be designed for adjusting coupling between resonators and for developing higher values of distributed capacitance [2].

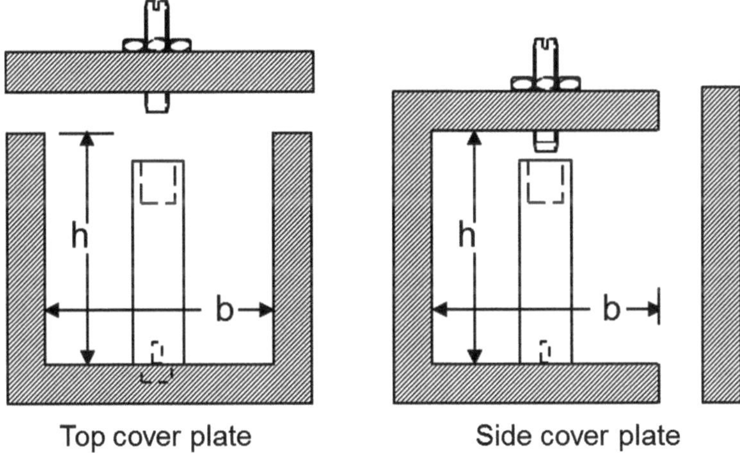

Fig. 7.2 Combline filter with two designs of cover plates

7.2.3 Cover Plate

Cavity block is designed either with a top cover plate perpendicular to resonators or with a side cover plate parallel to resonators as in Fig. 7.2 which shows the cross of a combline filter. The thickness of top cover plate is designed for 4 to 5 mm to facilitate the mounting of tuning and coupling screws. Lower thickness is adequate for side cover plate. Designing cavity block with side cover plate assists 'cut and try' method for marginally reducing the width of cavity for adjusting the bandwidth of filters. The contact area between cover plate and cavity block is a source of RF path discontinuity. Poor contact of the cover plate with cavity block results in increased losses. It is recommended that the screws for mounting cover plates to cavity block should not be spaced more than 10 to 15 mm to ensure good RF grounding.

7.2.4 Other Piece Parts

Input and output coaxial connectors are usually SMA or TNC or N connector series. The connectors with extended dielectric equal to the wall thickness of cavity block are used to avoid discontinuities in input/output impedance. Feed through terminals are used for the surface mounting of filters. Appendix-3 could be referred for additional information about coaxial connectors with extended dielectric and feed through terminals. Tuning and coupling screws are designed with 0.5 mm pitch for the precise tuning of cavity filters. Lock nuts prevents

Table 7.1 Sample specification of a combline RF filter

Electrical characteristics	Customer limit	Internal limit
Pass band centre frequency, fo	1 GHz	1 GHz
Bandwidth	80 MHz	85 MHz[a]
Insertion loss	2 dB max.	1.8 dB max.
Return loss	20 dB min.	22 dB min.
Rejection at (fo $\pm$ 100) MHz	60 dB min.	65 dB min.[b]
Pass band ripple (L_{ar})	0.4 dB max.	0.2 dB max.
Operating temperature	0–50 °C	0–50 °C
Input/output connectors	SMA (f)	SMA (f)

For calculating the combine band pass filter parameters

[a] Recommended 80 MHz as bandwidth for design calculations. 85 MHz to be ensured while tuning the filter

[b] Recommended 70 dB as rejection at (fo $\pm$ 100) MHz for design calculations. 65 dB to be ensured while tuning the filter

mechanical disturbance to the tuning and coupling screws. Fasteners are used for mounting resonators, cover plate and the coaxial connectors.

7.2.5 *Materials and Finish of Parts*

The materials and finish of the piece parts of filter are important for ensuring satisfactory RF performance. Cavity blocks fabricated with hard grade aluminium to prevent thread failures in tuning and coupling screws. Cavity block and cover plate are silver plated and the thickness of plating is designed between 3 and 15 μ depending on the frequency of application of filters. Brass is used for fabricating resonators, coupling probes, tuning screws, coupling screws and lock nuts and the parts are silver plated for 3 to 5 μ. Passivated stainless steel fasteners are recommended so that parts are tightened with high torque to ensure proper RF grounding.

7.3 Illustrated Design of Combline Filter

The design of Tchebyscheff combline band pass filter is illustrated with the sample electrical specification, shown in Table 7.1. Both customer and internal specification limits are shown.

Internal limits shown in Table 7.1 are used for the design of the combline filter. The design equations of filter are applied for combline filters with rectangular resonators. The rectangular resonators are modified to cylindrical resonators by approximations.

7.3.1 *Computational Steps for Combline Filter*

Combline filter design involves converting the electrical specifications into the mechanical dimensions of the parts of the filter. The mechanical dimensions that need to be computed are:

b, the width of the cavity.
l, the length of the cavity.
h, the height of the cavity.
L, the length of the rectangular resonator.
w, the width and t, thickness of the rectangular resonator.
s, the spacing between the resonators.
The diameter and the length of tuning and coupling screws.

In addition to the mechanical dimensions, the value of distributed capacitance, C_j^s, for tuning resonators and the number of resonators are also determined.

The filter design involves many intermediate computations for determining the values of mechanical design parameters. The steps in the computations are:

1. Determining the number of filter sections

    (a) Calculate ω, ratio of bandwidth to Centre frequency
    (b) Calculate ω'/ω_1', low pass to band pass transformation factor
    (c) Calculate ε

        i. $\varepsilon = [\mathrm{antilog}_{10}\ (L_{Ar}/10)] - 1$
        ii. L_{Ar} is the ripple in pass bandpass band

    (d) Calculate the number of filter sections, n

2. Determining resonator length
3. Determining the width and thickness of rectangular resonators

    (a) Compute g-values (element values of proto-type filters)
    (b) Compute G_T/Y_A and $J_{j,j+1}/Y_A$ (see Sect. 7.3.4.3)
    (c) Compute normalised capacitance/unit length (line and ground)
    (d) Compute normalised capacitance/unit length (adjacent resonators)
    (e) Obtain (s/b) and (C'_{fe}/ε) from graphs (see Sect. 7.3.4.6)
    (f) Computerisation of the graphs
    (g) Calculate normalised width of resonators
    (h) Decide the width of cavity

4. Approximating rectangular resonators to cylindrical resonators
5. Calculating spacing between resonators
6. Calculating the length of cavity
7. Calculating the capacitance, C_j^s, for tuning resonators
8. Calculating the height of cavity and realisation of C_j^s

7.3.2 Number of Filter Sections

The required data for determining the number filter sections is reproduced from the filter specifications shown in Table 7.1.

$$\text{fo} = 1000 \text{ MHz}$$

$$f_1, \text{ lower pass band edge frequency} = \text{fo} - \text{Bandwidth}/2$$
$$= 1000 - 80/2 = 960 \text{ MHz}$$

$$f_2, \text{ upper pass band edge frequency} = \text{fo} + \text{Bandwidth}/2$$
$$= 1000 + 80/2 = 1040 \text{ MHz}$$

7.3.2.1 Calculate ω

$$\omega = (f_2 - f_1)/\text{fo} = (1040 - 960)/1000 = 0.08$$

7.3.2.2 Calculate ω'/ω_1'

$$f, \text{ upper rejection frequency} = \text{fo} + 100 \text{ MHz}$$
$$= 1000 + 100 = 1100 \text{ MHz}$$

$$\omega'/\omega_1' = (2/\omega) * [(f - \text{fo}) / \text{fo}]$$
$$= (2/0.08) * [(1100 - 1000) / 1000] = 2.5$$

7.3.2.3 Calculate ε

$$L_{Ar}, \text{ pass band ripple} = 0.2 \text{ dB}$$

$$\varepsilon = [\text{antilog}_{10}(L_{Ar}/10)] - 1$$
$$= [\text{antilog}_{10}(0.2/10)] - 1 = 0.047$$

7.3.2.4 Calculate the Number of Filter Sections, n

L_A rejection at upper rejection frequency, 1100 MHz = 70 dB.
$n = \cosh^{-1}[\text{SQRT} \{(10^\wedge (L_A/10)-1/\varepsilon\}] / \cosh^{-1}(\omega'/\omega_1')$

n = $\cosh^{-1}$[SQRT {(10^ (75/10)$-$1/0.047}] / $\cosh^{-1}$(2.5) = 6.56
n is rounded up to the next higher integer.
The number of filter sections, n = 7

7.3.3 Resonator Length

λo/8 resonators are used in combline filters.

$$\text{Resonator length} = \lambda o/8 = 300000 \,/\, (8/\text{fo})$$
$$= 300000 \,/\, (8/1000) = 37.5 \text{ mm}$$

7.3.4 Width and Thickness of Rectangular Resonators

7.3.4.1 Compute (b_j/Y_A) for j = 1 to n

θo, electrical length of the λo/8 resonator = 360/8 = 45° = 0.786 radians

Y_{aj}, the resonator line admittance, is chosen as 1/(70 Ω) to ensure high unloaded Q-factor[1] for resonators. The normalised resonator line admittance, (Y_{aj}/Y_A), is calculated as:

$$\left.\frac{Y_{aj}}{Y_A}\right|_{j=1 \text{ to } n} = \frac{1/70}{1/50} = 0.714$$

$$\left.\frac{b_j}{Y_A}\right|_{j=1 \text{ to } n} = \frac{Y_{aj}}{Y_A}\left(\frac{\cot\theta_0 + \theta_0 \csc^2\theta_0}{2}\right)$$

$$\left.\frac{b_j}{Y_A}\right|_{j=1 \text{ to } n} = 0.714\left(\frac{\cot 45 + 0.786\ \csc^2 45}{2}\right) = 0.918$$

7.3.4.2 Compute g-values

The number of the g-values that need to be computed is related to the number of resonators. For n resonators, g_o to g_{n+1} values are computed. For n = 7, g_o to g_8 values are computed. Though the g-values are tabulated for ready use in the Microwave Filter design book [1], the equations given in the Design Book are explained for computerisation.

$$g_0 = 1$$

For other g-values, the intermediate values, β, γ, a_k and b_k, are calculated.

$$\beta = \ln[\coth(L_{Ar}/17.37)]$$
$$L_{Ar}, \text{ Pass band ripple } = 0.2$$
$$\beta = \ln[\coth(0.2/17.37)] = 4.464$$
$$\gamma = \sinh[\beta/(2n)] = \sinh[4.464/(2*7)] = 0.324$$
$$a_k = \sin[\Pi(2k-1)/(2\,n)] \quad k = 1, 2, \ldots, n$$

The calculated a_k values are shown in Table 7.2.

$$b_k = \gamma 2 + \sin^2(k\Pi/n) \quad k = 1, 2, \ldots, n$$

The calculated b_k values are shown in Table 7.3.

$$g_1 = 2\,a_1/\gamma$$
$$g_k = (4\,a_{k-1}a_k) \,/\, (b_{k-1}g_{k-1}) \quad k = 2, 3, \ldots, n$$
$$g_{n+1} = 1 \text{ for n odd}$$
$$= \coth^2(\beta/4) \text{ for n even}$$

The calculated g-values are shown in Table 7.4.

7.3.4.3 Compute G_T/Y_A and $J_{j,j+1}/Y_A$

The set of design equations for computing G_T/Y_A and $J_{j,j+1}/Y_A$ are:

$$\frac{G_{T1}}{Y_A} = \frac{\omega\left(\frac{b_1}{Y_A}\right)}{g_0 g_1 \omega_1'}$$

$$\left.\frac{J_{j,j+1}}{Y_A}\right|_{j=1 \text{ to } n-1} = \frac{\omega}{\omega_1'} \sqrt{\left[\frac{\left(\frac{b_j}{Y_A}\right)\left(\frac{b_{j+1}}{Y_A}\right)}{g_j g_{j+1}}\right]}$$

$$\frac{G_{Tn}}{Y_A} = \frac{\omega\left(\frac{b_n}{Y_A}\right)}{g_n g_{n+1} \omega_1'}$$

ω_1', normalized prototype parameter in radians = 1

From Sect. 7.3.2.1, $\omega = 0.08$

From Sect. 7.3.4.1, $\left.\dfrac{b_j}{Y_A}\right|_{j=1 \text{ to } n} = 0.918$

The calculated values of G_{Tn}/Y_A and $J_{(j,\ j+1)}/Y_A$ are shown in Table 7.5.

Table 7.2 Calculated values of a_k

K	a_k
1	0.2226
2	0.6237
3	0.9012
4	1.0000
5	0.9006
6	0.6227
7	0.2214

Table 7.3 Calculated values of b_k

K	b_k
1	0.5390
2	0.8870
3	1.0800
4	1.0797
5	0.8862
6	0.5379
7	0.1037

Table 7.4 Calculated g-values

g-values	Calculated value for 0.2 dB ripple
g_0	1.0000
g_1	1.3741
g_2	1.3776
g_3	2.2774
g_4	1.4993
g_5	2.2772
g_6	1.3771
g_7	1.3695
g_8	1.0000

Table 7.5 Calculated values of G_{Tn}/Y_A and $J_{(j,\ j+1)}/Y_A$

G_{T1}/Y_A	0.0534
$J_{1,2}/Y_A$	0.0534
$J_{2,3}/Y_A$	0.0415
$J_{3,4}/Y_A$	0.0397
$J_{4,5}/Y_A$	0.0397
$J_{5,6}/Y_A$	0.0415
$J_{6,7}/Y_A$	0.0535
G_{T7}/Y_A	0.0536

7.3.4.4 Normalized Capacitance/Unit Length (Line and Ground)

For calculating normalised capacitances per unit length between the resonators and ground, two additional end resonators need to be assumed as per filter Microwave

Filter Design Book [1]. For the combline filter with n = 7, the additional end resonators are Resonator-0 and Resonator-8. The set of equations for calculating the normalised capacitances are:

$$\frac{C_0}{\varepsilon} = \frac{376.7 \, Y_A}{\sqrt{\varepsilon_r}} \left(1 - \sqrt{\frac{G_{T1}}{Y_A}} \right)$$

$$\frac{C_1}{\varepsilon} = \frac{376.7 \, Y_A}{\sqrt{\varepsilon_r}} \left[\left(\frac{Y_{a1}}{Y_A} \right) - 1 + \left(\frac{G_{T1}}{Y_A} \right) - \left(\frac{J_{1,2}}{Y_A} \right) \tan \theta_0 \right] + \frac{C_0}{\varepsilon}$$

$$\left. \frac{C_1}{\varepsilon} \right|_{j=2 \text{ to } n-1} = \frac{376.7 \, Y_A}{\sqrt{\varepsilon_r}} \left[\left(\frac{Y_{a1}}{Y_A} \right) - \left(\frac{J_{j-1,j}}{Y_A} \right) \tan \theta_0 - \left(\frac{J_{j,j+1}}{Y_A} \right) \tan \theta_0 \right]$$

$$\frac{C_n}{\varepsilon} = \frac{376.7 \, Y_A}{\sqrt{\varepsilon_r}} \left[\left(\frac{Y_{an}}{Y_A} \right) - 1 + \left(\frac{G_{Tn}}{Y_A} \right) - \left(\frac{J_{n-1,n}}{Y_A} \right) \tan \theta_0 \right] + \frac{C_{n+1}}{\varepsilon}$$

$$\frac{C_{n+1}}{\varepsilon} = \frac{376.7 \, Y_A}{\sqrt{\varepsilon_r}} \left(1 - \sqrt{\frac{G_{Tn}}{Y_A}} \right)$$

ε_r: Relative dielectric constant of air = 1

From Sect. 7.3.4.1:

$$\theta_o = 0.786 \text{ radians}$$

$$Y_A = (1/50) \text{ mhos}$$

$$\left. \frac{Y_{aj}}{Y_A} \right|_{j=1 \text{ to } n} = 0.714$$

$$\left. \frac{C_0}{\varepsilon} \right| = \frac{376.7(1/50)}{\sqrt{1}} \left(1 - \sqrt{0.0534} \right) = 5.7925$$

$$\left. \frac{C_1}{\varepsilon} \right| = \frac{376.7(1/50)}{\sqrt{1}} [(0.714) - 1 + (0.0534) - (0.0534) \tan 45] + 5.7925$$

$$= 3.6402$$

Similarly, the values of C_2/ε to C_8/ε are calculated and the values are shown in Table 7.6.

7.3.4.5 Normalised Capacitance/Unit Length (Adjacent Resonators)

The normalised capacitances per unit length are calculated between adjacent resonators. C_{01} is the capacitance between resonators 0 and 1. C_{12} is the capacitance between resonators 1 and 2 and so on. The set of equations for the calculation are:

Table 7.6 Calculated values of C_j/ε

C_j/ε	Value
C_0/ε	5.7925
C_1/ε	3.6402
C_2/ε	4.6666
C_3/ε	4.7694
C_4/ε	4.7823
C_5/ε	4.7693
C_6/ε	4.6658
C_7/ε	3.6378
C_8/ε	5.7894

$$\frac{C_{0,1}}{\varepsilon} = \frac{376.7\ Y_A}{\sqrt{\varepsilon_r}} - \frac{C_0}{\varepsilon} = \frac{376.7\ (1/50)}{\sqrt{1}} - 5.7925 = 1.74$$

$$\left.\frac{C_{j,j+1}}{\varepsilon}\right|_{j=1\ \text{to}\ n-1} = \left(\frac{376.7\ Y_A}{\sqrt{\varepsilon_r}}\right)\left(\frac{J_{j,j+1}}{Y_A}\right)\tan\theta_0$$

$$\frac{C_{1,2}}{\varepsilon} = \frac{376.7\ Y_A}{\sqrt{\varepsilon_r}}\left(\frac{J_{1,2}}{Y_A}\right)\tan\theta_0 = \frac{376.7\ (1/50)}{\sqrt{1}}(0.0534)\tan 45 = 0.40$$

Similarly, $(C_{1,2}/\varepsilon)$, $(C_{2,3}/\varepsilon)$, $(C_{3,4}/\varepsilon)$, $(C_{4,5}/\varepsilon)$, $(C_{5,6}/\varepsilon)$ and $(C_{6,7}/\varepsilon)$ are calculated.

$$\frac{C_{n,n+1}}{\varepsilon} = \frac{376.7\ Y_A}{\sqrt{\varepsilon_r}} - \frac{C_{n+1}}{\varepsilon}$$

$$\frac{C_{7,8}}{\varepsilon} = \frac{376.7\ Y_A}{\sqrt{\varepsilon_r}} - \frac{C_8}{\varepsilon} = \frac{376.7\ (1/50)}{\sqrt{1}} - 5.7895 = 1.74$$

The calculated values of $(C_{j,j+1}/\varepsilon)$ are shown in Table 7.7.

7.3.4.6   Obtain (s/b) and (C'_{fe}/ε) From Graphs

The values of (s/b) and (C'_{fe}/ε) are obtained from the design graphs shown in Fig. 7.3. The x-axis of the graph represents (s/b) and the values of the ratio range from 0 to 1.5. The y-axis of the graph represents two capacitance parameters, $(\Delta C/\varepsilon)$ and (C'_{fe}/ε), and the values of capacitance range from 0.01 to 10 pF. The figure contains two sets of graphs, namely, $(\Delta C/\varepsilon)$ vs (s/b) and (C'_{fe}/ε) vs (s/b). Each set has six graphs for six different values of (t/b), which ranges from 0.0 to 0.8.

s   Spacing between rectangular resonators
t   Thickness of rectangular resonators
b   Width of cavity block

Table 7.7 Calculated values of $(C_{j,j+1}/\varepsilon)$

$C_{j,j+1}/\varepsilon$	Value
$C_{0,1}/\varepsilon$	1.74
$C_{1,2}/\varepsilon$	0.40
$C_{2,3}/\varepsilon$	0.31
$C_{3,4}/\varepsilon$	0.30
$C_{4,5}/\varepsilon$	0.30
$C_{5,6}/\varepsilon$	0.31
$C_{6,7}/\varepsilon$	0.40
$C_{7,8}/\varepsilon$	1.74

$$\frac{\Delta c}{\varepsilon} = \text{Normalised capacitance/unit length(adjacent resonators)} = \frac{c_{j,\,j+1}}{\varepsilon}$$

The procedure to obtain the values of (s/b) and (C'_{fe}/ε):

- Select the desired value for (t/b)
- Use the value of $[(C_{j,j+1})/\varepsilon]$, enter the graph, $(\Delta C/\varepsilon)$ vs (s/b), for the selected value of (t/b) and obtain the ratio, (s/b)
- Use the ratio of (s/b), enter the graph, (C'_{fe}/ε) vs (s/b), for the same value of (t/b) and obtain the value of (C'_{fe}/ε)

7.3.4.7 Obtaining $(s_{0,\,1}/b)$ from $(C_{0,\,1}/\varepsilon)$

The procedure to obtain the values of (s/b) and (C'_{fe}/ε) is illustrated for obtaining the values of $(s_{0,1}/b)$ from $(C_{0,1}/\varepsilon)$ assuming the value of (t/b) is 0.2.

$$C_{0,\,1}/\varepsilon = 1.74 \text{ pF}$$

The value, 1.74, cuts the x-axis of the graph, $(\Delta C/\varepsilon)$ vs (s/b), at 0.19 i.e. $(s_{0,1}/b) = 0.19$. The values of $(s_{j,j+1}/b)$ are obtained from the graph for all the values of $C_{j,j+1}$ and they are shown in Table 7.8.

7.3.4.8 Obtain $[(C'_{fe})_{j,j+1}/\varepsilon)]$ From $(s_{0,1}/b)$

Obtaining the values $[(C'_{fe})_{j,j+1}/\varepsilon)]$ from $(s_{0,1}/b)$ is illustrated for the same value of (t/b) i.e. 0.2.

$$s_{0,1}/b = 0.19$$

The value, 0.19, cuts the y-axis of the graph, (C'_{fe}/ε) vs (s/b), at 0.21 i.e. $[(C'_{fe})_{0,1}/\varepsilon)] = 0.21$. The values of $[(C'_{fe})_{j,j+1}/\varepsilon)]$ are obtained from the graph for all the values of $(s_{j,j+1}/b)$ and they are shown in Table 7.9.

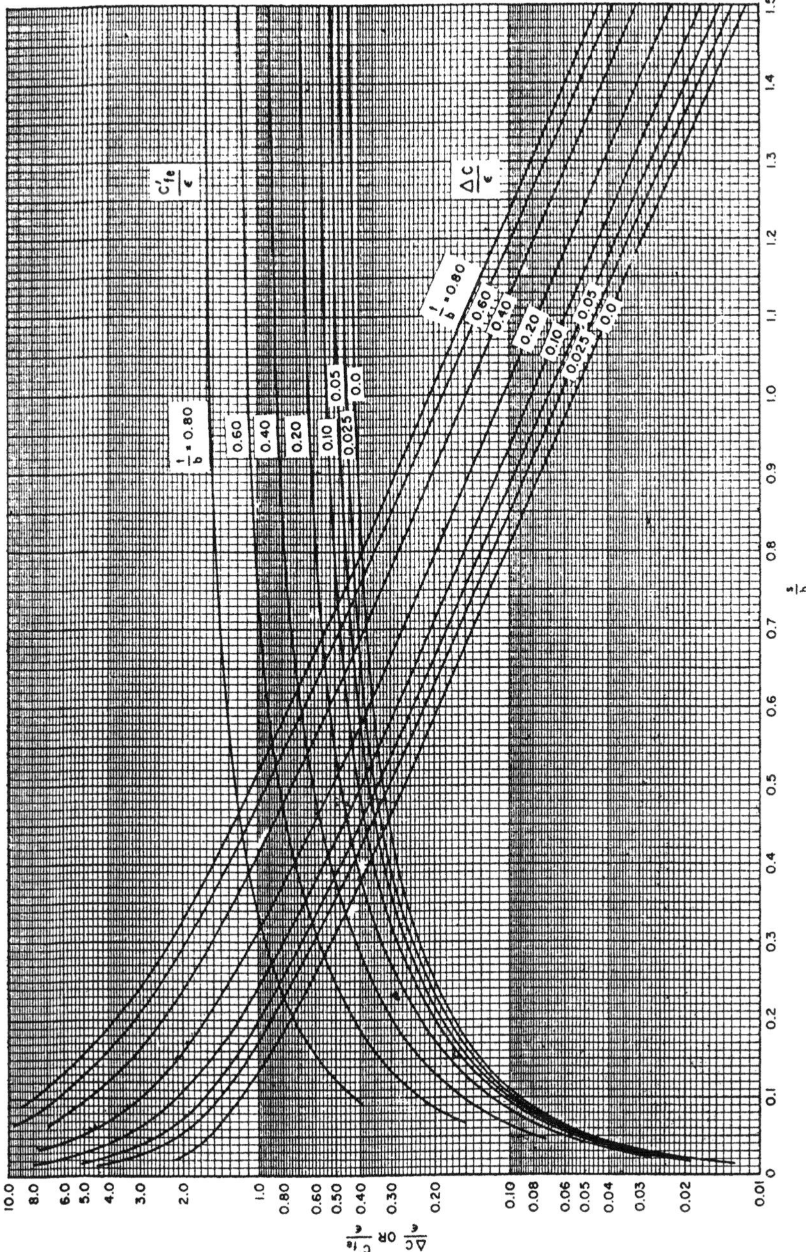

Fig. 7.3 Normalized even-mode fringing capacitance c'_{fe}/ε and interbar capacitance $\Delta C/\varepsilon$ for coupled rectangular bars (Reproduced by permission from [1], MA: Artech House, Inc., 1980. © 1980 by Artech House, Inc)

Table 7.8 Values of $s_{j,j+1}/b$ for the values of $C_{j,j+1}$

$C_{j,j+1}$	$s_{j,j+1}/b$
$C_{0,1}/\varepsilon = 1.74$	$s_{0,1}/b = 0.19$
$C_{1,2}/\varepsilon = 0.40$	$s_{1,2}/b = 0.58$
$C_{2,3}/\varepsilon = 0.31$	$s_{2,3}/b = 0.66$
$C_{3,4}/\varepsilon = 0.30$	$s_{3,4}/b = 0.68$
$C_{4,5}/\varepsilon = 0.30$	$s_{4,5}/b = 0.68$
$C_{5,6}/\varepsilon = 0.31$	$s_{5,6}/b = 0.66$
$C_{6,7}/\varepsilon = 0.40$	$s_{6,7}/b = 0.58$
$C_{7,8}/\varepsilon = 1.74$	$s_{7,8}/b = 0.19$

Table 7.9 Values of $[(C'_{fe})_{j,j+1}/\varepsilon)]$ for the values of $s_{j,j+1}/b$

$s_{j,j+1}/b$	$[(C'_{fe})_{j,j+1}/\varepsilon)]$
$s_{0,1}/b = 0.19$	$[(C'_{fe})_{0,1}/\varepsilon)] = 0.21$
$s_{1,2}/b = 0.59$	$[(C'_{fe})_{1,2}/\varepsilon)] = 0.51$
$s_{2,3}/b = 0.68$	$[(C'_{fe})_{2,3}/\varepsilon)] = 0.55$
$s_{3,4}/b = 0.69$	$[(C'_{fe})_{3,4}/\varepsilon)] = 0.55$
$s_{4,5}/b = 0.69$	$[(C'_{fe})_{4,5}/\varepsilon)] = 0.55$
$s_{5,6}/b = 0.68$	$[(C'_{fe})_{5,6}/\varepsilon)] = 0.55$
$s_{6,7}/b = 0.59$	$[(C'_{fe})_{6,7}/\varepsilon)] = 0.51$
$s_{7,8}/b = 0.19$	$[(C'_{fe})_{7,8}/\varepsilon)] = 0.21$

7.3.4.9 Computerization of the Graphs

The need to refer the design graph shown in Fig. 7.3 for obtaining the values of $(s_{j,j+1}/b)$ and $[(C'_{fe})_{j,j+1}/\varepsilon)]$ is eliminated to facilitate the development of combline filter design software. Polynomial expressions are fitted to the design graphs using computer oriented numerical techniques. Lagrange interpolation technique [3] is applied to fit polynomial expressions for both the sets of graphs, $(\Delta C/\varepsilon)$ vs (s/b) and (C'_{fe}/ε) vs (s/b). The application of the interpolation technique is explained using a second degree polynomial,

$$f(x) = c_1(x - x_2)(x - x_3) + c_2(x - x_1)(x - x_3) + c_3(x - x_1)(x - x_2)$$

7.3.4.10 Polynomials for the Graph, $[(\Delta C)_{j,j+1}/\varepsilon]$ i.e. $C_{j,j+1}/\varepsilon$ Vs $s_{j,j+1}/b$

For better accuracy, the graph for $t/b = 0.2$, is divided into six segments as shown below so that the error is less than 5 % and the fitting a second degree polynomial is explained for the segment-6.

Segment-1	Segment-2	Segment-3	Segment-4	Segment-5	Segment-6
0.022–0.195	0.195–0.73	0.73–1.6	1.6–3.55	3.55–5.2	5.2–7

Table 7.10 Verification of the polynomial for segment-6

$x = [(\Delta C)_{j,j+1}/\varepsilon] = C_{j,j+1}/\varepsilon$	$f(x) = s_{j,j+1}/b$ from polynomial)	$f(x) = s_{j,j+1}/b$ (from the graph)	Error (%)
6.4	0.0407	0.040	1.7
6.0	0.0435	0.043	1.1
5.4	0.0483	0.048	0.5
5.0	0.0518	0.052	0.4

For the segment-6, (5.2 to 7):

The values of x are obtained for three values of f (x) within the range of segment-6 from the graph and they are tabulated below.

$x = [(\Delta C)_{j,j+1}/\varepsilon] = C_{j,j+1}/\varepsilon$ $f(x) = s_{j,j+1}/b$	$x_1 = 5.2$ $f(x_1) = 0.05$	$x_2 = 5.8$ $f(x_2) = 0.045$	$x_3 = 7$ $f(x_3) = 0.037$

$$f(x) = c_1(x - x_2)(x - x_3) + c_2(x - x_1)(x - x_3) + c_3(x - x_1)(x - x_2)$$

The value of c_1 is determined by substituting $x = x_1$. The value of c_2 is determined by substituting $x = x_2$. The value of c_3 is determined by substituting $x = x_3$.

$$C_1 = \frac{f(x_1)}{(x_1 - x_2)(x_1 - x_3)} = \frac{0.050}{(5.2 - 5.8)(5.2 - 7.0)} = 0.0463$$

$$C_2 = \frac{f(x_2)}{(x_2 - x_1)(x_2 - x_3)} = \frac{0.045}{(5.8 - 5.2)(5.8 - 7.0)} = -0.0625$$

$$C_3 = \frac{f(x_3)}{(x_3 - x_1)(x_3 - x_2)} = \frac{0.037}{(7.0 - 5.2)(7.0 - 5.8)} = 0.0171$$

$$f(x) = 0.0463(x - 5.8)(x - 7) - 0.0625(x - 5.2)(x - 7) + 0.0171(x - 5.2)(x - 5.8)$$

The fitting of the polynomial is verified and the results are shown in Table 7.10.

The second degree Lagrange polynomials for the graph, $(\Delta C/\varepsilon)$ vs (s/b) for t/b = 0.2 are determined they are shown in Table 7.11.

7.3.4.11 Polynomials for the Graph, $s_{j,j+1}/b$ Vs $[(C'_{fe})_{j,j+1}/\varepsilon]$

The second degree Lagrange polynomials for the graph, $[(C'_{fe})_{j,j+1}/\varepsilon)]$ for t/b = 0.2 are shown in Table 7.12. The Lagrange polynomials for other values of t/b could also be determined.

Table 7.11 Polynomials for the graph, $C_{j,j+1}/\varepsilon$ Vs $s_{j,j+1}/b$ for $t/b = 0.2$

Segment, $[(\Delta C)/\varepsilon]$	Polynomial for the segment, $[(\Delta C)/\varepsilon]$
0.022–0.195	$f(x) = 51.381(x - 0.105)(x - 0.022) - 133.869(x - 0.195)$ $(x - 0.022) + 104.464(x - 0.195)(x - 0.105)$
0.195–0.73	$f(x) = 2.1061(x - 0.375)(x - 0.195) - 9.3897(x - 0.73)$ $(x - 0.195) + 8.3074(x - 0.73)(x - 0.375)$
0.73–1.60	$f(x) = 0.4105(x - 1.04)(x - 0.73) - 1.7281(x - 1.6)$ $(x - 0.73) + 1.4831(x - 1.6)(x - 1.04)$
1.60–3.55	$f(x) = 0.0311(x - 2.23)(x - 1.6) - 0.1684(x - 3.55)$ $(x - 1.6) + 0.1628(x - 3.55)(x - 2.23)$
3.55–5.20	$f(x) = 0.0233(x - 3.9)(x - 3.55) - 0.1538(x - 5.2)$ $(x - 3.55) + 0.1385(x - 5.2)(x - 3.55)$
5.20–7.00	$f(x) = 0.0463(x - 5.8)(x - 7) - 0.0625(x - 5.2)$ $(x - 7) + 0.0171(x - 5.2)(x - 5.8)$

Table 7.12 Polynomials for the graph, $s_{j,j+1}/b$ Vs $[(C'_{fe})_{j,j+1}/\varepsilon)]$ for $t/b = 0.2$

Segment, s/b	Polynomial for the segment, s/b
0.02 to 0.4	$f(x) = 0.3947(x - 0.2)(x - 0.4) - 6.2778(x - 0.02)(x - 0.4)$ $+ 5.3947(x - 0.02)(x - 0.2)$
0.4 to 1.0	$f(x) = 2.2778(x - 0.7)(x - 1) - 6.2222(x - 0.4)(x - 1) + 3.5(x - 0.4)(x - 0.7)$
1.0 to 1.5	$f(x) = 5.04(x - 1.25)(x - 1.5) - 10.48(x - 1)(x - 1.5) + 5.44(x - 1)(x - 1.25)$

7.3.4.12 Calculate Normalized Width of Rectangular Resonators

The additional end resonators, Resonator-0 and Resonator-8 are not required for the final realisation of the combline filter. The normalised width of the seven rectangular resonators, w_j/b, is obtained from the expression,

$$\left.\frac{w_j}{b}\right|_{j=1 \text{ to } 7} = \frac{1}{2}\left(1 - \frac{t}{b}\right)\left[\frac{1}{2}\left(\frac{C_j}{\varepsilon}\right) - \left\{\frac{(C'_{fe})_{j-1,j}}{\varepsilon}\right\} - \left\{\left(\frac{(C'_{fe})_{j,j+1}}{\varepsilon}\right)\right\}\right]$$

The normalised width of Resonator-1 is calculated from the expression for w_1/b.

$$\left.\frac{w_1}{b}\right. = \frac{1}{2}\left(1 - \frac{t}{b}\right)\left[\frac{1}{2}\left(\frac{C_1}{\varepsilon}\right) - \left\{\frac{(C'_{fe})_{0,1}}{\varepsilon}\right\} - \left\{\frac{(C'_{fe})_{1,2}}{\varepsilon}\right\}\right]$$

$$\left.\frac{w_1}{b}\right. = \frac{1}{2}(1 - 0.2)\left[\frac{1}{2}(3.6402) - \{0.21\} - \{0.51\}\right] = 0.44$$

Similarly, the normalised widths of Resonator-2 through Resonator-7 are calculated and the values are shown in Table 7.13.

Table 7.13 Normalised width of rectangular resonators, w_j/b

w_j/b	Value
w_1/b	0.44
w_2/b	0.51
w_3/b	0.51
w_4/b	0.52
w_5/b	0.51
w_6/b	0.51
w_7/b	0.44

Table 7.14 Width of rectangular resonators, w_j

N	w_j/b for $j = 1$ to n	w_j for $j = 1$ to n (mm)
1	0.44	6.16
2	0.51	7.13
3	0.51	7.19
4	0.52	7.23
5	0.51	7.19
6	0.51	7.13
7	0.44	6.16

7.3.4.13 Decide the Width of Cavity

The width of cavity of a combline filter has its implications on the Insertion loss and the size of the filter. Higher width results in large sized filter with lower insertion loss. Using past design data is important in deciding the diameter of cavity. Considering size and design data on similar cavity filters, a width (b) of 14 mm is assumed. Library of cavity dimensions linked to the characteristics of filters are developed and updated over a period of time.

7.3.4.14 Calculate the Thickness and Width of Rectangular Resonators

Width of the cavity, $b = 14$ mm

Thickness of rectangular resonators, $t = (t/b) \times b = 0.2 \times 14 = 2.8$ mm
Width of the rectangular resonator-1, $w_1 = (w_1/b)b = 0.44 \times 14 = 6.16$ mm
The width of the other rectangular resonators is calculated and the values are shown in Table 7.14.

7.3.5 Rectangular to Cylindrical Resonators

The rectangular resonators are approximated to cylindrical resonators to enhance the produceability of the resonators. Equating the circumference of resonators

Table 7.15 Diameters of cylindrical resonators, d_j

N	w_j for $j = 1$ to n (mm)	d_j for $j = 1$ to n (mm)
1	6.16	5.70
2	7.13	6.32
3	7.19	6.36
4	7.23	6.38
5	7.19	6.36
6	7.13	6.31
7	6.16	5.70

Table 7.16 Calculated values of spacing between cylindrical resonators, $s_{j,j+1}$

n	$s_{j,j+1}/b$ for $j = 1$ to n $- 1$	$s_{j,j+1}$ for $j = 1$ to n $- 1$ (mm)
1	$s_{1,2}/b = 0.58$	8.12
2	$s_{2,3}/b = 0.66$	9.24
3	$s_{3,4}/b = 0.68$	9.52
4	$s_{4,5}/b = 0.68$	9.52
5	$s_{5,6}/b = 0.66$	9.24
6	$s_{6,7}/b = 0.58$	8.12

minimally disturbs the calculated distributed capacitance values compared to equating the surface area of resonators. Equating the circumference of both the types of resonators,

$$\Pi \, d_j = 2 \, (w_j + t)$$

The diameters of the cylindrical resonators are calculated using the above expression and they are shown in Table 7.15.

Another approximation is done by averaging the diameters to have symmetrical resonators in the combline filter and it does not affect complying with the electrical specifications of the combline filter.

$$d = (5.70 + 6.32 + 6.36 + 6.38 + 6.36 + 6.31 + 5.70)/7 = 6.2 \text{ mm}$$

7.3.6 Spacing Between Resonators

Spacing between resonators $= (s_{j,j+1}/b) \times b$

Spacing between Resonator-1 and Resonator-2 $= 0.58 \times 14 = 8.12$ mm

The calculated values of spacing between the resonators are shown in Table 7.16.

It is recommended to reduce the calculated values of the spacing between resonators by 5 % to minimise 'cut and try' efforts. It does not lead to any errors as

Table 7.17 Final values of spacing between cylindrical resonators $s_{j,j+1}$

n	$s_{j,j+1}$ for $j = 1$ to $n - 1$ (mm)
1	7.7
2	8.8
3	9.0
4	9.0
5	8.8
6	7.7

the filter could be tuned to achieve the specified performance. The final spacing between the resonators is shown in Table 7.17.

7.3.7 Length of Cavity

The length of cavity is calculated from the diameter of the resonators and the spacing between the resonators. Further, a gap of 5 mm is required to couple the end resonators (Resonator-1 and Resonator-7) to the input/output coaxial connectors.

l, cavity length $= 2 \times$ End gap $+ 7 \times$ resonator dia $+$ sum of spacing.

Sum of resonator spacing $= 7.7 + 8.8 + 9.0 + 9.0 + 8.8 + 7.7 = 51.0$ mm.

l, the length of the cavity $= 2 \times 5 + 7 \times 6.2 + 51.0 = 104.4$ mm.

7.3.8 Capacitance for Tuning Resonators

Each resonator is tuned to resonate at the centre frequency of pass band using the distributed capacitor, C_j^s, developed at the free end of the resonator. The distributed capacitance for tuning the resonators is calculated using the expression,

$$C_j^s\bigg|_{j=1 \text{ to } n} = (Y_A)\left(\frac{Y_{aj}}{Y_A}\right)\left(\frac{\cot \theta_0}{\omega_0}\right)$$

$$(Y_{aj}/Y_A) = 0.714$$

$$\omega_0 = 2\prod f_0 = 2*\prod *1000 * 10^6 = 0.0062832 * 10^{12}$$

$$C_j^s\bigg|_{j=1 \text{ to } 7} = (1/50)(0.714)\left(\frac{\cot 45}{0.0062832 * 10^{12}}\right) = 2.27\text{pF}$$

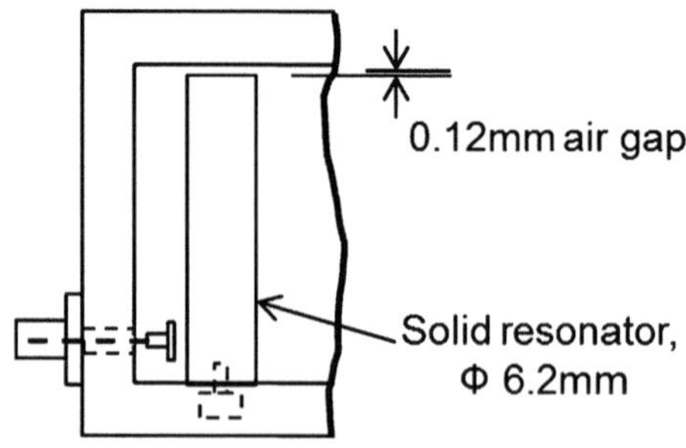

Fig. 7.4 Tuning capacitance at the air gap

7.3.9 *Height of Cavity*

The height of a combline filter is decided by the length of resonator and the air gap to realise the distributed capacitance, C_j^s. The fundamental expression for capacitance is used for calculating the capacitance.

C	Capacitance of parallel plate (Farads) = Area $\times$ ε/distance
ε	Permittivity = $\varepsilon_r \times \varepsilon_v$
ε_r	Relative permittivity or relative dielectric constant
ε_v	Permittivity of space = 8.854×10^{-12} F/m
C (pF)	= $0.008854 \times \varepsilon_r \times$ Area/gap
Area	Cross sectional area of the resonator in mm^2
Distance	gap between the resonator and cavity block in mm

For the resonator with $\Phi 6.2$ mm, a gap of 0.12 mm gives the required $C_j^s = 2.27$ and it is shown in Fig. 7.4. However, maintaining such a narrow gap between the resonators and the wall of the cavity block affects both the produceability and reliability of the filter. It is practically impossible to tune the resonators with the air gap of 0.12 mm. Also, the filter performance is degraded in the specified operating temperature range and transportation vibration could cause detuning of the filter. Hence, 1 mm gap between the resonators and the wall of the cavity block is assumed.

$$\text{h, height of the cavity} = 1\text{ mm} + \text{Length of resonator}$$
$$= 1 + 37.5 = 38.5\text{ mm}$$

C_j^s is realised using hollow resonator and tuning screw as shown in Fig. 7.5. Hollow resonator has a blind hole drilled at the free end. A tuning screw is threaded from the cavity block into the hollow end of the resonator. Distributed capacitance exists between the tuning screw and the hollow free end of resonator. C_j^s is calculated using the expression for the capacitance of coaxial line [4].

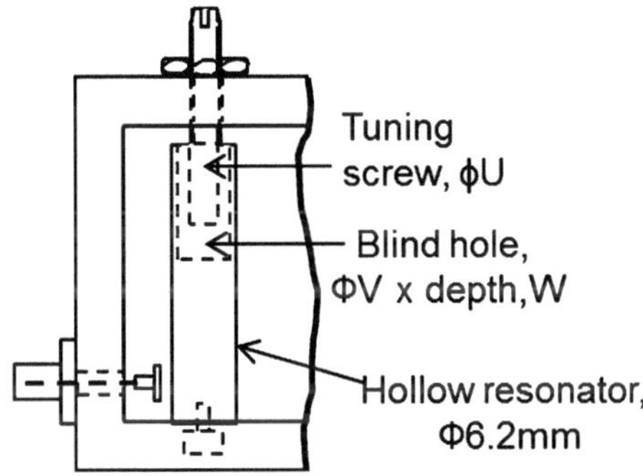

Fig. 7.5 Tuning capacitance using hollow resonator

Table 7.18 Mechanical design parameters for the combline filter of Table 7.1

Mechanical design parameter	Designed value (mm)
b, the width of the cavity	14
l, the length of the cavity	104.4
h, the height of the cavity	38.5
L, the length of the cylindrical resonator	37.5
d, diameter of cylindrical resonator	6.2
Number of resonators (with blind hole of Φ5 mm for a depth of 12 mm at the free end)	7
s, the spacing between the resonators	$s_{1,2} = 7.7$ $s_{2,3} = 8.8$ $s_{3,4} = 9.0$ $s_{4,5} = 9.0$ $s_{5,6} = 8.8$ $s_{6,7} = 7.7$
Tuning screws	Φ4
Coupling screws	Φ4

$$C = 0.02414 \times \varepsilon_T / \log(m/n) \text{ pF/mm}$$

$$C_j^s = W \times 0.02414 \times \varepsilon_T / \log(V/U) \text{ pF}$$

ε_r = Relative dielectric constant
V   Diameter of blind hole in mm
U   Diameter of screw in mm
W   Depth of blind hole in mm

Assume that a blind hole of Φ5 mm is drilled at the free end of resonator, Φ6.2 mm. screw, Φ4 mm, with a travel to a depth of 10 mm into the resonator develops approximately 2.5pF, ignoring the variations caused by the threads of the tuning screw. The travel depth is finalised as 12 mm with a margin of 2 mm. Coupling screw, Φ4 mm, is decided considering standardisation of fasteners. The initial length of the tuning and coupling screws is assumed to be 15 mm for the

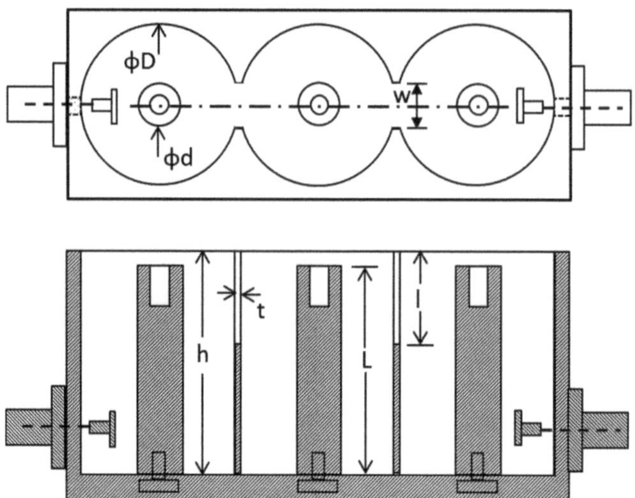

Fig. 7.6 Design dimensions of iris-coupled band pass filter

design of the filter and the exact value is decided at the end of tuning. The mechanical design dimensions to satisfy the electrical specifications of the combline filter are summarised in Table 7.18.

7.4 Iris-Coupled Band Pass Filters

7.4.1 Structure and Parts

The general structure of iris-coupled cavity band pass filter is shown in Fig. 7.6 indicating the mechanical dimensions that need to be computed for the piece parts and irises of the filter. The number of cavities is limited to three in Fig. 7.6 for simplicity. Like combline filter, iris-coupled filter also has cavity block, resonators, cover plate, coupling probes, tuning/coupling screws, lock nuts, input/output coaxial connectors and fasteners. Each circular cavity is coupled to adjacent cavity through an iris. Iris-coupled filters are always designed with top cover plates. The materials and finish of the parts are same as those of combline filters.

7.5 Illustrated Design of Iris-Coupled Filters

The design of Tchebyscheff iris-coupled band pass filter is illustrated with the sample electrical specification, shown in Table 7.19. Both customer and internal specification limits are shown.

Table 7.19 Sample specification of a Iris-coupled RF filter

Electrical characteristics	Customer limit	Internal limit
Pass band centre frequency, fo	2.15 GHz	2.15 GHz
Bandwidth	10 MHz	12 MHz[a]
Insertion loss	1.5 dB max.	1.35 dB max.
Return loss	20 dB min.	22 dB min.
Rejection at (fo $\pm$ 150) MHz	85 dB min.	90 dB min.[b]
Pass band ripple	0.4 dB max.	0.2 dB max.
Operating temperature	0–50 °C	0–50 °C
Input/output connectors	SMA (f)	SMA (f)

For calculating the iris-coupled band pass filter parameters

[a] Recommended 95 dB as rejection at (fo $\pm$ 150) MHz for design calculations. 90 dB to be ensured while tuning the filter

[b] Recommended 10 MHz as bandwidth for design calculations. 12 MHz to be ensured while tuning the filter

7.5.1 Computational Steps for Filter Design Parameters

The electrical specifications of iris-coupled filters are converted into filter design parameters. The design parameters are the mechanical dimensions of the parts of the filter. The mechanical dimensions that need to be computed are:

D, the diameter of cavity
h, the height of the cavity block.
L, the length of resonator.
d, the diameter of resonator
The width, height & thickness of irises.

The diameter and the length of tuning and coupling screws.
Number of filter sections i.e. the number of resonators
The design calculations are done using the mid-range frequency i.e. the centre frequency of the band pass filter and hence the suffix, m, is used for the notations of various parameters. Filter design involves many intermediate computations for determining the values of mechanical design parameters. The steps in the computations are:

1. Determining the number of filter sections
2. Computing cavity/
3. Computing the height, width and thickness of irises

- Calculate g-values
- Calculate coupling reactance, $(X_{j,j+1})_m$
- Calculate initial magnetic polarizability of the irises, $(M_{j,j+1})$

- Initial width and length of irises from graph
- Calculate compensated magnetic polarizability of the irises, $(M_{j,j+1})_{comp}$
- Final width and length of irises from graph

4. Deciding the diameter and the length of & coupling screws.

7.5.2 Number of Filter Sections

The procedure for determining the number of filter sections is exactly same as that explained for combline band pass filter.

$$fo = 2150 \text{ MHz}$$

$$f_1, \text{ lower pass band cut - off frequency } = 2145 \text{ MHz}$$

$$f_2, \text{ upper pass band cut - off frequency } = 2155 \text{ MHz}$$

$$\omega = (f_2 - f_1) / fo = (2155 - 2145) / 2150 = 0.00466$$

$$f, \text{ upper rejection frequency} = fo + 150 \text{ MHz}$$
$$= 2150 + 150 = 2300 \text{ MHz}$$

$$\omega'/\omega'_1 = (2/\omega) * [(f - fo) / fo]$$
$$= (2/0.00466) * [(2300 - 2150)/2150] = 30$$

$$L_{Ar}, \text{ pass band ripple } = 0.2 \text{ dB for design}$$

$$\varepsilon = [\text{ antilog}_{10}(L_{Ar}/10)] - 1$$
$$= [\text{ antilog}_{10}(0.2/10)] - 1 = 0.047$$

L_A rejection at upper rejection frequency, 2300 MHz = 95 dB
$n = \cosh^{-1}[\text{SQRT } \{(10^\wedge (L_A/10) - 1/\varepsilon\}]/\cosh^{-1}(\omega'/\omega_1')$
$n = \cosh^{-1}[\text{SQRT } \{(10^\wedge (95/10) - 1/0.047\}]/\cosh^{-1}(30) = 3.21$
n should be rounded up to the next higher integer
The number of filter sections, n = 4

7.5.3 Diameter and Height of Cavity and Resonators

Coaxial cavities are designed for nominal characteristic impedance (Z_o) of 76 Ω for obtaining optimum resonator Q-factor [1].

$$Z_0 = \frac{60}{\sqrt{\varepsilon_r}} \ln \frac{D}{d}$$

D Inner diameter of the coaxial cavity
d Outer diameter of the resonators

The inner diameter of the coaxial cavity has its implications on the Insertion loss and the size of the filter. Higher diameter results in large sized filter with lower insertion loss. Using past design data is important in deciding the diameter of cavity. Considering size and design data on similar cavity filters, cavity diameter of 35 mm for the iris-coupled filter is assumed. Library of cavity dimensions linked to the characteristics of filters are developed and updated over a period of time.

ε_r = 1 as the dielectric medium inside the filter is air
D   Inner diameter of cavity = 35 mm (1.575 in.)
d   Diameter of resonators = 9.9 mm

Iris-coupled filter uses $\lambda/4$ resonators.
Mid-band frequency = 2150 MHz
λ_m, wavelength of the mid-band frequency = 139.5 mm

L, Length of resonator = $\lambda_m/4$ = 35 mm

Assuming a clearance of 2 mm at the free end of resonators for tuning,

h, height of cavity = L + 2 mm = 37 mm

7.5.4 Width, Height and Thickness of Irises

Rectangular irises are normally used with their lengths parallel to resonators. The length (l), width (w) and thickness (t) of the irises are shown in Fig. 7.5. The length represents the depth of irises from the top of cavity. Designing irises parallel to resonators supports the easy application of cut and try method for adjusting bandwidth. As the number of filter sections (resonators) is four, there are three irises in the iris-coupled band pass filter. As the graphs and expressions in the Microwave Filter Design Book [1] use inches, all calculations are done in inches and the final values of width, height and thickness of irises are converted and expressed in mm.

7.5.4.1 Calculate g-Values

The method of calculating g-values is exactly same as that explained for combline filters. The calculated g-values for the three cavity iris-coupled band pass filter are shown in Table 7.20.

Table 7.20 g-values for the Iris-coupled RF filter

g_i	Calculated g-values for 0.2 dB ripple
g_0	1.0000
g_1	1.3034
g_2	1.2842
g_3	1.9763
g_4	0.8452
g_5	1.5386

Table 7.21 Values of $(X_{j,j+1})_m$

$(X_{j,j+1})_m$	Value
$(X_{1,2})_m$	0.6669
$(X_{2,3})_m$	0.5416
$(X_{3,4})_m$	0.6676

7.5.4.2 Calculate Coupling Reactance, $(X_{j,j+1})_m$

Using the values given in the Microwave Filter Design Book [1] for $(\theta_A)_m$ and $(\theta)_m$ are quite adequate for computing the values of the coupling reactance between cavities.

$$(\theta_A)_m = 0.9675 \text{ radians}$$

$$(\theta)_m = 0.8534 \text{ radians}$$

$$\left.(X_{j,j+1})_m\right|_{j=1 \text{ to } n-1} = \frac{\pi Z_0 \omega_m}{4 \cos \theta_m^2 \, \omega_1' \sqrt{g_j \, g_{j+1}}}$$

$$\omega_m = \text{Bandwidth} / f_0 = 10/2150 = 0.0047$$

$$Z_0 = 76 \, \Omega$$

$$\omega_1' = 1$$

Coupling reactance between the cavities, $(X_{1,\,2})_m$, $(X_{2,\,3})_m$ and $(X_{3,\,4})_m$, are calculated and the values are shown in Table 7.21.

7.5.4.3 Calculate Initial Magnetic Polarizabilities of the Irises, $(M_{j,j+1})$

Magnetic polarizabilities, $(M_{j,j+1})$, of the irises are calculated using the expression,

$$(M_{j,j+1}) = \frac{\lambda_m D^2 (X_{j,j+1})_m}{60}$$

Table 7.22 Values of $(M_{j,j+1})$

$(M_{j,j+1})$	Value in inches
$(M_{1,2})$	0.029
$(M_{2,3})$	0.024
$(M_{3,4})$	0.029

λ_m wavelength of mid-band frequency (f_o) in inches = 5.49
D cavity diameter in inches = 1.575

The calculated values of magnetic polarizabilities are shown in Table 7.22.

7.5.4.4 Initial Width and Length of Irises From Graph

The magnetic polarizability graph, shown in Fig. 7.7 is referred for obtaining the width and length of rectangular irises. The graph relates the variables, w/l and M/l^3, where w is the width of irises, l is the length of irises and M is the magnetic polarizability.

The graph relates three types of slots i.e. irises. Two sets of values for the variables are read from the graph for the rectangular irises and they are indicated below.

For w/l = 0.2, the value of M/l^3 is 0.09
For w/l = 1.0, the value of M/l^3 is 0.2575

The relationship between the two variables is approximately linear in the range of the values (0.09 and 0.2575) mentioned above and hence a linear expression is fitted for the graph. The values of width (w) and length (l) of the irises are chosen such that the ratio of w/l lies between 0.2 and 1.0. The magnetic polarizability expression relating the two generalised variables for the linear range of the graph is given below:

$$\left(\frac{w}{l_{j,j+1}}\right) = 4.75 \left[\frac{(M_{j,j+1})}{(l^3)_{j,j+1}}\right] - 0.2425$$

In the design of iris-coupled band pass filters, the width of all the irises are usually kept same and hence only the length of the irises needs to be determined. The width of irises is assumed to be 12.7 mm i.e. 0.5 in. and it is appropriate for the cavity diameter of 35 mm. The lengths of the irises, $l_{1,2}$, $l_{2,3}$ and $l_{3,4}$, are determined from the magnetic polarizability expression using the values of $(M_{j,j+1})$.

$$\left(\frac{w}{l_{j,j+1}}\right) = 4.75 \left[\frac{(M_{j,j+1})}{(l^3)_{j,j+1}}\right] - 0.2425$$

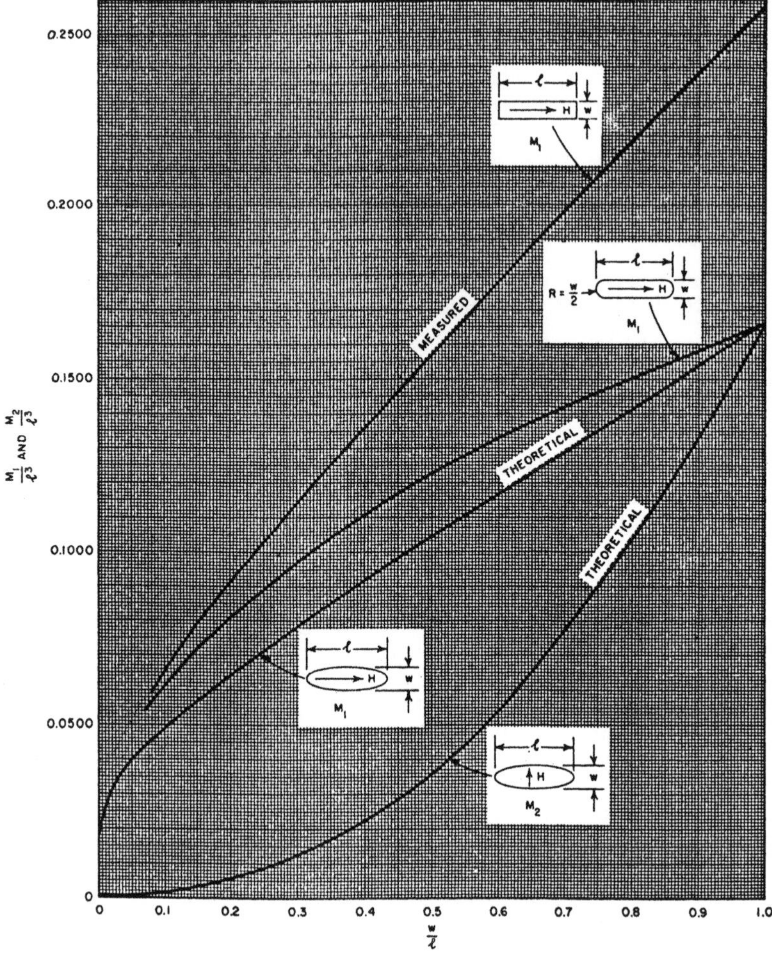

Fig. 7.7 Magnetic polarizabilities of rectangular, rounded-end, and elliptical slots (Reproduced by permission from [1], MA: Artech House, Inc., 1980. © 1980 by Artech House, Inc)

$M_{1,2}$ = 0.029

w   = 0.5 inches for all the three irises

$$\left(\frac{0.5}{l_{1,2}}\right) = 4.75\left[\frac{(0.029)}{\left(l^3\right)_{1,2}}\right] - 0.2425$$

The cubic equation is solved for obtaining the values of $l_{1,\,2}$ by using excel tool or software or standard calculators.

$l_{1,2}$ = 0.47 in.

   Similarly, the other two lengths of the irises are determined.
$l_{2,3}$ = 0.43 in.
$l_{3,4}$ = 0.47 in.

7.5.4.5 Calculate Compensated Magnetic Polarizabilities of the Irises, $(M_{j,j+1})_{comp}$

The lengths of the irises determined above are approximate and their accuracies are improved by computing compensated magnetic polarizabilities of the irises and recalculating the lengths of the irises. The expression for the calculation is:

$$\left(M_{j,j+1}\right)_{comp} = \left(M_{j,j+1}\right)\left[1 - \left(2\,l_{j,j+1}/\lambda_m\right)^2\right]10^{\left[1.36\ t/l_{j,j+1}\right]}\sqrt{\left[\left(1-\left(2l_{j,j+1}/\lambda_m\right)\right)^2\right]}$$

   For calculating compensated magnetic polarizabilities of the irises, the thickness (t) of irises is required. Higher coupling between cavities is achieved with lower values of thickness. The width of the irises could be decreased by reducing the thickness of irises. Considering machinability requirements, the thickness of irises is usually set between 1 and 2 mm. For the iris-coupled filter, the thickness (t) of irises is set to 1.6 mm = 0.063 in.
   Thickness of irises, t = 1.6 mm = 0.063 in.
   The calculated values of the compensated polarizabilities of the irises are shown in Table 7.23.

7.5.4.6 Final Width and Length of Irises From Graph

Final values for the lengths of the irises are calculated using the same magnetic polarizability expression except that $(M_{j,j+1})$ is replaced by $(M_{j,j+1})_{comp}$. The final width of irises remains same i.e. 12.7 mm i.e. 0.5 in. The final values of the lengths of the irises, $(l_{1,2})_{comp}$, $(l_{2,3})_{comp}$ and $(l_{3,4})_{comp}$ are determined from the magnetic polarizability expression using the values of $(M_{j,j+1})_{comp}$.

$$\left[\frac{w}{\left(l_{j,j+1}\right)_{comp}}\right] = 4.75\left[\frac{M_{j,j+1}}{\left\{\left(1^3\right)_{j,j+1}\right\}_{comp}}\right] - 0.2425$$

$(M_{1,2})_{comp,}$ = 0.042
w   = 0.5 in. for all the three irises

Table 7.23 Values of $(M_{j,j+1})_{comp}$

$(M_{j,j+1})$ inches	$(M_{j,j+1})_{comp}$ inches
$(M_{1,2}) = 0.029$	$(M_{1,2})_{comp} = 0.042$
$(M_{2,3}) = 0.024$	$(M_{2,3})_{comp} = 0.036$
$(M_{3,4}) = 0.029$	$(M_{3,4})_{comp} = 0.042$

Table 7.24 Mechanical design parameters for the Iris-coupled filter of Table 7.19

Mechanical design parameter	Designed value (mm)
D, the diameter of each cavity	35
h, the height of the cavity	37
L, the length of the cylindrical resonator	35
d, the diameter of resonator	9.9
Number of resonators (with a blind hole of $\Phi 5$ mm for a depth of 12 mm at one end)	4
Iris dimensions	width, w = 12.7 $l_{1,2} = 14.3$ $l_{2,3} = 13.3$ $l_{3,4} = 14.3$
Tuning screws	$\Phi 4$
Coupling screws	$\Phi 4$

$$\left[\frac{0.5}{(l_{1,2})_{comp}}\right] = 4.75 \left[\frac{0.029}{\left\{(l^3)_{1,2}\right\}_{comp}}\right] - 0.2425$$

$$\left(l_{1,2}\right)_{comp} = 0.56 \text{ inches} = 14.3 \text{ mm}$$

$$\left(l_{2,3}\right)_{comp} = 0.52 \text{ inches} = 13.3 \text{ mm}$$

$$\left(l_{3,4}\right)_{comp} = 0.56 \text{ inches} = 14.3 \text{ mm}$$

The ratio of the final calculated values of the width and length of irises is between 0.2 and 1.0, the linear range of the magnetic polarizability graph, Fig. 5.10.4 (a) of the Microwave Filter Design Book [1]. Distributed capacitance at the free end of resonators is realised by having blind hole in resonators as explained in combline filters. 'cut and try' method cannot be eliminated but it could be greatly minimised, saving design cycle time. For example, using $\Phi 4$ mm tuning screws and resonators with $\Phi 5$ mm blind holes is part of minimising 'cut and try' efforts. Increasing the diameters of blind holes could be implemented easily. More information for minimising 'cut and try' efforts is available in Chap. 8, Practical Expertise for Filter Design. The mechanical design dimensions to meet the electrical specifications of the iris-coupled filter are shown in Table 7.24.

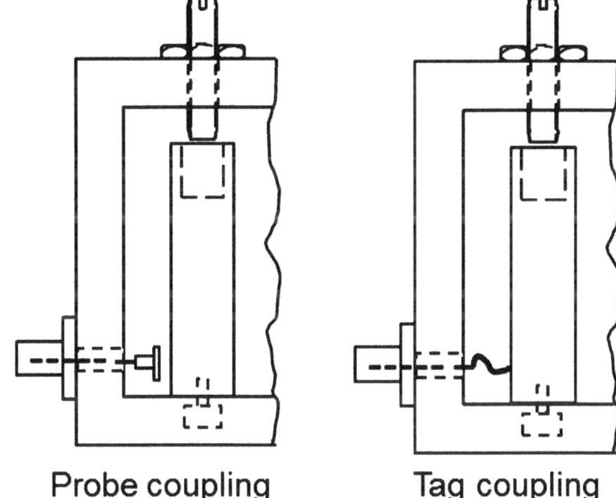

Fig. 7.8 Types of to resonators

7.6 End Coupling in Cavity Band Pass Filters

End coupling refers to the method of interconnecting the end resonators to the input/output coaxial RF transmission line coaxial connectors of microwave cavity filters. also matches the higher resonator line impedance, used in the design of the filters, to the impedance of input/output coaxial connectors, 50 Ω i.e. it serves as impedance transformation [5]. Air-looping or wire connection method is used for end coupling and they are applicable for both combline and iris-coupled cavity filters. Probe coupling is another terminology for the air-looping method of coupling. Probe coupling and tag coupling is shown in Fig. 7.8.

7.6.1 Probe Coupling

In the air-looping method, there is no physical contact between the centre contacts of coaxial connectors and the end resonators. Probes are used to air-couple the end resonators near their short-circuited ends (grounded ends) to the centre contacts of input/output coaxial connectors. In the probe method of end coupling, circular probes are soldered to the centre contact of the input/output coaxial connectors. The probe design is such that its position could be axially moved adjusting the air gap between the end resonator and the input/output connector, i.e. adjusting the coupling to the end resonators. The diameter of the probe could also be varied for coupling.

7.6.2 Tag Coupling

In the wire connection method, there is physical contact between the centre contacts of coaxial connectors and the end resonators. Silver plated copper wire tags are soldered between the centre contacts of the input/output coaxial connectors and the end resonators near their short-circuited ends (grounded ends). The tags are not straight piece of wires but they are in bent forms. The shape of the wire tag is adjusted for varying the coupling between the end resonators and the input/output connectors. The height of tag soldering point from the grounded end of the resonators also alters the coupling. Annealed copper wire, $\Phi 0.8$ mm, is used for the wire tags to facilitate the adjustment of their shapes. The precautions to be observed during the adjustment of tag coupling are explained in Chap. 8 (see Sect. 7.5).

7.7 Tuning of Cavity Band Pass Filters

Microwave cavity filters, when assembled; do not meet the specified electrical specifications. Each filter needs to be tuned for satisfying the specifications. Although the tuning of cavity filters seems to be difficult and time consuming, operators could be trained to become experts in tuning. Vector Network Analyser (VNA) is ideal and most suited for the tuning of cavity filters as it assists novice tuners to become experts faster. General information for tuning in the VNA set-up is presented in Chap. 8.

7.7.1 Adjustment Available for Tuning

In cavity band pass filters, the following adjustments are available for tuning:

1. Tuning screws for tuning resonators to the centre frequency of filter
2. Coupling screws to adjust the coupling between resonators for adjusting the bandwidth of filter
3. Probe adjustment in filters with probe coupling for adjusting return loss, if required:

   - Adjusting diameter of the probe
   - Adjusting gap between the probe and the end resonators

4. Tag adjustment in filters with tag coupling for adjusting return loss, if required:

   - Adjusting shape of tag
   - Adjusting height at which the tags are soldered from the grounded end of resonators

Probe or tag coupling is adjusted if the required return loss of filter could not be achieved by adjusting tuning and coupling screws. Improving return loss improves the insertion loss of the filter also. Detailed design output documents are prepared to eliminate or minimise the need for the adjustment of probe or tag during the manufacturing of filters. For high frequency cavity band pass filters, innovative design for the fine adjustment of tag coupling is required for satisfying the return loss and insertion loss of filters.

7.7.2 Tuning Procedure

VNA has two ports, namely, source port and detector port. Source port provides input signal for RF filters. Detector port functions as a load and monitors the detected signal at the output of filters. After the assembly of cavity band pass filter, the input of the filter is connected to the source port and the output of the filter is connected to the detector port of VNA. The centre frequency, span and other settings in VNA are selected as per the general information in Chap. 8. The VNA displays two traces. One trace displays the transmission characteristic and the other displays the reflection characteristic of the filter. Initially, random values of insertion loss and return loss are displayed by the VNA traces in the specified pass band of filter.

The first step in tuning process is to obtain the approximate pass band of the filter. The approximate pass band is obtained by tuning each resonator to the centre frequency of the filter. The tuning is started with the end resonator at the input. Then the end resonator at the output is tuned. Middle resonators are tuned after tuning the end resonators. While tuning the next adjacent resonator, the tuning of previously tuned resonator is slightly disturbed due to coupling between them. The previously tuned resonator is tuned again and the tuning is continued with remaining resonators. In the process of tuning resonators, the insertion loss and return loss performances improve in the pass band. Tuning is an iteration process and it is continued until the best possible insertion loss and return loss is realised in the pass band.

The next step in the tuning process is to adjust the coupling between resonators for controlling the bandwidth of filter. Higher bandwidth reduces insertion loss in the pass band but it poses difficulties in satisfying rejection performance limits. Hence, coupling between resonators need to be adjusted balancing bandwidth, insertion loss and rejection performance limits. In the process of coupling adjustment, the resonators get slightly de-tuned and hence they are also re-tuned.

If the return loss and insertion loss of filter could not be met by adjusting tuning and coupling screws, then the probe or tag coupling to end resonators is adjusted as described in Sect. 7.7.1. After adjusting the probe or tag coupling, tuning and coupling screws are fine-tuned for complying with the specifications of the filter as per internal specification limits. The tuning and coupling screws are then locked using lock nuts.

7.7.3 Tuning of Filters in Frequency and Time Domains

VNA supports the tuning of filters both in frequency domain and in time domain. Expert tuners prefer frequency domain as the performance of filters could be verified simultaneously while tuning. It is better for novice tuners to use time domain for the tuning of filters as it has the advantage of learning to become expert tuners. Basically, VNA displays all the discontinuities present between its two ports in time domain. The advantages of tuning RF filters in time domain are:

(1) The discontinuities in resonators i.e. the status of tuning of all resonators is displayed simultaneously. For example, if a cavity filter has five resonators, the status of tuning of the five resonators is displayed. Operators could tune the resonators for best possible return loss using the status display

(2) The iteration process of tuning could be done easily. While tuning a resonator or adjusting coupling between resonators, the disturbance to the tuning of previously tuned resonators is displayed in VNA. Novice tuners could use the display and retune the disturbed resonators.

VNA is set to frequency domain from time domain for viewing the performance of filters. Application notes, AN 1287-8 and AN 1287-10 of Agilent could be referred for the tuning of filters in time domain. Cavity filters that are not complying with the specified electrical characteristics in operation, would require 'cut and try' as described in Chap. 8. Information on cross-coupling of filters in Chap. 8 could also be considered.

References

1. Matthei GL, Young Leo, Jones EMT (1980) Microwave filters, impedance-matching networks, and coupling structures. Artech House, Norwood
2. Guohua Shen, Djuradj Budimir (2002) Novel resonator structures for combline filter applications. Microwave conference, 32 European, pp 1–3
3. Rajaraman V (1985) Computer oriented numerical techniques. Prentice Hall International
4. John D (1965) Ryder, networks, lines and fields. Prentice-Hall, Inc., New Jercy
5. Amir Borji, Dan Busuioc, Safavi-Naeini S, Chaudhuri SK (2002) ANN and EM based models for fast and accurate modeling of excitation loops in combline-type filters. IEEE MTT-S, pp 2105–2108

Chapter 8
Practical Expertise for Filter Design

Abstract Useful suggestions and recommendations for the design and manufacturing of RF filters based on learning and literature information are presented. Preventing the degradation of the reliability of piece parts, tuning, 'cut and try' method, cross coupling for steep roll down filters and developing filter design software for the illustrated tutorials are some of the important topics covered in the chapter.

8.1 Opportunities for Learning

Parts and processes fail for various reasons during the design and manufacturing of RF filters. Failures are opportunities for new learning provided analysis is done for understanding the root causes of failures. This chapter presents useful suggestions and recommendations for the design and manufacturing of RF filters based on learning and literature information.

The topics that are briefly presented are:

1. Visual examination of assembled PCBs
2. Preservation and Handling of silver plated parts
3. Precautions for chip ceramic capacitors
4. Centre contact retention of coaxial connectors
5. Mechanical design
6. Tuning of RF filters
7. 'Cut and try' method
8. Cross coupling for steep roll down RF filters
9. Air gap between resonators and tuning screws
10. Duplexers
11. Developing filter design software.

D. Natarajan, *A Practical Design of Lumped, Semi-Lumped and Microwave Cavity Filters*, Lecture Notes in Electrical Engineering,
DOI: 10.1007/978-3-642-32861-9_8, © Springer-Verlag Berlin Heidelberg 2013

8.2 Visual Examination of Assembled PCBs

Printed circuit boards (PCBs) packaged with inductors and capacitors are used in lumped RF filters. PCBs are cleaned with appropriate solvent to remove flux residues after soldering. RF filters are high value modules and are used in equipments, requiring high reliability. Flux residues are trapped especially near chip capacitors and they cause performance degradation with time affecting reliability. Hence, it is economical to subject the filters for 100 % visual examination under $10\times$ magnification to ensure that PCBs are free from flux residues before tuning.

8.3 Preservation of Silver Plated Parts

Silver plated copper wires are used for winding inductors for lumped and semi-lumped filters. Almost all the piece parts of microwave cavity filters are silver plated. Silver plated parts are passivated after silver plating to prevent tarnishing of the parts. If tarnished silver plated parts are used in RF filters, it results in higher insertion loss. Hence, it is necessary to take additional precautions during the preservation and handling of the parts. Precautions such as storage of parts in sulphur free environment and using gloves for handling should be followed.

8.4 Precautions for Chip Ceramic Capacitors

Chip ceramic capacitors are used in low power lumped RF filters. The performance of the capacitors could be degraded during soldering and post cleaning operations, affecting the reliability of RF filters in equipments. Manufacturers of chip capacitors recommend the use of silver solder to prevent the leaching of capacitor terminations. It is also necessary to complete the soldering operation for the capacitors in 3 to 5 s approximately.

After the soldering of capacitors, flux residues are cleaned using the solvent, isopropyl alcohol or trichloroethane. The temperature of capacitor is approximately 250 °C immediately after soldering. The cleaning solvents are highly volatile. If flux residue is cleaned with the solvent immediately after soldering, the solvent causes liquid thermal shock to the capacitor. Depending on the intensity of thermal shock, instant failure or developing micro-cracks in the ceramic layers of chip capacitor could occur, affecting the reliability of RF filters. Hence, post cleaning operation is done after the capacitors cool down to near ambient temperature but before the flux residue hardens.

8.5 Centre Contact Retention in Coaxial Connectors

Panel mount coaxial connector sockets or plugs are used in RF filters. The centre contacts of the connectors are captivated to prevent their longitudinal and rotational movements by the manufacturers of coaxial connectors. Catalogues of connector manufacturers indicate that connectors with mechanically or epoxy captivated centre contacts are available. If the captivation of centre contacts is disturbed in finished RF filters, tuning of the filters is disturbed and the filters fail to comply with the specified performance requirements of the filters.

The captivation of centre contacts in coaxial connectors is usually not affected in lumped and semi-lumped RF filters, but it could be degraded in microwave cavity filters. The centre contacts of coaxial connectors are coupled to end resonators using either probes or tags. The shape of the tags and the point of soldering the tags to end resonators are adjusted in tuning operation for achieving specified return loss. In probe coupling, the gap between the probe and the end resonators is adjusted by de-soldering the probes from the centre contacts of coaxial connectors. Such adjustment operations on tags or probes cannot be eliminated in the tuning of filters. However, the operations apply thermal and mechanical stresses on the captivation of the centre contacts of coaxial connectors. The stresses degrade the reliability of the connectors and the RF filters might fail in equipments. Replacement of coaxial connectors in completed RF filters is an expensive task.

Simple precautions could be observed by RF filter designers.

1. Silver plated annealed copper wire is used for tag terminations. The diameter of the wire is decided considering flexibility and vibration resistance. The use of lower diameter wire is flexible but the tag tuning adjustment might be disturbed by transportation vibration. Wire diameter between 0.6 and 0.8 mm could be considered by filter designers.

2. If tag coupling needs to be adjusted after de-soldering, allow the cooling of centre contact to ambient temperature before the mechanical adjustment. This prevents the application of combined thermal and mechanical stresses to the centre contact. The epoxy captivated centre contacts are more vulnerable to the combined stresses compared to mechanically captivated centre contacts.

3. In probe coupling, axial force should not be applied to the probe for adjusting it until the solder melts completely. If mechanical force is applied to the centre contact of coaxial connectors under thermal stress, the captivation of the centre contact is disturbed.

4. Adjustment of tag coupling should be done gently using proper tools that would avoid jerks or slipping. Jerks or slipping applies excessive mechanical shock stresses on the captivation of centre contacts.

5. Filter designers should provide detailed documents for the fabrication of pre-shaped tags and the location of soldering point from the grounded ends of resonators for tag coupling. For probe coupling, the gap between the probe and end resonators should be documented. The documentation

minimises both mechanical and thermal stresses on the centre contacts of coaxial connectors.

Additionally, the input and output coaxial connectors of completed RF filters could be tested for Centre contact retention for axial load. The use of special tools and set-up, recommended by the manufacturers of coaxial connectors, are mandatory for conducting the test to prevent the testing induced failures on the coaxial connectors.

8.6 Mechanical Designs

A good percentage of mechanical design co-exists with the electronic design of RF filters especially with tubular filters and microwave cavity filters. Innovation and creativity are essential in the design of mechanical parts of the filters. Innovative mechanical design significantly contributes for the produceability, reliability and performance of the filters. Satisfactory RF grounding in all the mechanical interfaces of cavity filters and design of fasteners for tuning/locking of filters are examples that might require innovative mechanical design. Knowledge in mechanical design is also needed in deciding the tolerances of the dimensions of parts to avoid conflicting dimensions in the drawings of piece parts. Electronic engineers should develop mechanical design expertise needed for the design RF filters to minimise time and cost.

8.7 Tuning of Filters

VNA is preferred for the tuning and testing of RF filters. The required frequency window is set in the display of VNA. The span and the centre frequency decide the frequency window settings. The frequency window should be such that all the specified characteristics of RF filter is displayed without changing the frequency window settings during tuning. The reference for insertion loss and return loss displays is set to 0 dB for convenience.

Tuning of filters requires precision adapters and interconnection cables. If assembled cables are used, they are checked for return loss to avoid errors in the tuning of RF filters. Appendix-3 could be referred for more information about Standard cables, assembled cables and precision adapters. VNA is calibrated with the test cables and adapters that are used for the tuning of filters. The manuals of VNA could be referred for the initial set-up calibration of VNA.

RF filters are firmly held for tuning. If clamps are used for holding RF filter, they should not compress the cavities of the filter. Proper tools are necessary for tuning and locking of filters. Care is exercised to prevent mechanical loading on input/output connectors of filters and other connector interfaces. Mechanical loading damages the female contacts of connectors and adapters.

8.8 'Cut and Try' Method

Microwave cavity filters that are fabricated as per design computations are tuned to comply with the specified requirements of the filters. If the requirements could not be met, the filters require some rework to satisfy the performance requirements. The rework method is known as 'cut and try' [1] method. Examples rework are changes in the width of cavity or in the spacing between resonators or in the diameter of resonators. In the literature method of developing RF filters, 'cut and try' efforts cannot be eliminated but could be minimised with experience.

'Cut and try' efforts could be minimised in cavity filters by marginally modifying the filter parameters that are computed.

1. Reducing the spacing between resonators by 5 %
2. Increasing the diameters of resonators by 5 %
3. Designing filters with side cover plate to facilitate the milling of cavity block for reducing the width of cavity. The side cover plate could be changed to top cover plate for the manufacturing of filters
4. Lower iris dimensions.

8.9 Cross Coupling for Steep Roll Down Filters

Certain applications require RF filters having steep fall from pass band. An example of such a requirement of RF filter is shown in Table 8.1.

Iris-coupled cavity band pass filter is more suitable for cross coupling. Normally, the coupling in an iris-coupled filter is serial between adjacent resonator sections i.e. cavity-1 is coupled to cavity-2; cavity-2 is coupled to cavity-3 and so on. With this normal serial coupling arrangement alone, the specified insertion loss in the pass band and steep roll down from pass band cannot be met. Hence, in addition to the serial coupling, cross coupling between resonator sections is designed to satisfy the steep fall requirements of iris-coupled filter. The cross coupling could be between cavity-2 and cavity-5 or between cavity-1 and cavity-4. Cross coupling reduces insertion loss at band pass band edges and results in steep roll down from pass band edge frequencies.

To begin with, the design parameters of iris-coupled filters are computed as per the procedure explained in Chap. 7. Normally, the resonator sections of a filter are arranged in a single row. The arrangement of single row filter topology is not suitable for cross coupling. Hence, the filter topology is decided to facilitate cross coupling between the required cavities. The literature mentioned in the references [1, 2] could be referred for designing the cross coupling between resonator sections.

Cross coupling is achieved by placing a short length of a semi-rigid coaxial cable between cavities. The construction of semi-rigid cable makes it more suitable for cross coupling. Semi-rigid coaxial cable has seamless copper tube as outer

Table 8.1 Specification of a RF filter with steep roll down

Parameter	Limit
Pass band	(995–1,005) MHz
Insertion loss	1 dB max.
Return loss	20 dB min.
Rejection at 985 and 1,015 MHz	50 dB min.
Operating temperature	0–50 °C
Input/output connectors	SMA (f)

conductor and copper clad steel silver plated inner conductor. PTFE is the dielectric of the cable. The cable is available with diameters, 0.086″, 0.141″ and 0.250″. Out of the three sizes, the semi-rigid cable with diameter, 0.141″ (RG402) is more convenient for cross coupling.

The outer conductor of semi-rigid cable is firmly held between the cavity block of filter and its cover plate by having suitable grooves on both of them. The depth of the groove should be sufficient to prevent excessive compression of the cable. Only the inner conductor of the cable protrudes into the cavities which have to be cross-coupled. The length and shape of the protrusion of the inner conductor of the coaxial cable decide the amount of cross coupling required for the steep roll down from the pass band edge frequencies of filter. Adjusting the length and shape of the inner conductor is part of filter tuning and it needs to be done patiently as it requires the removal of cover plate for adjustment. The cross coupling arrangement in an iris-coupled cavity band pass filter is shown in Fig. 8.1.

8.10 Air Gap Between Resonator and Tuning Screw

There is air gap between tuning screw and the free end of resonator and the air gap develops the required distributed capacitance for tuning resonators. Decreasing the gap develops higher values of distributed capacitance. However, decreasing the gap below acceptable value affects produceability, temperature performance and reliability of cavity filters.

Figure 8.2 shows only the resonator with its tuning screw of a microwave cavity filter. The diameter of the tuning screw is d and the diameter of the blind hole in the resonator is D.

The air gap, g, between the resonator and the tuning screw is given by,

$$g = (D - d)/2$$

Tuning screws and resonators are designed to be concentric but inherent variations in concentricity occur in machining operations. If the gap between resonator and tuning screw is small, the inherent concentricity variation causes short circuit between tuning screw and resonator during the tuning of filters and affects the reliability of filters during transportation and application. Maintaining a gap of 0.5 to 1.0 mm between resonators and tuning screws is recommended.

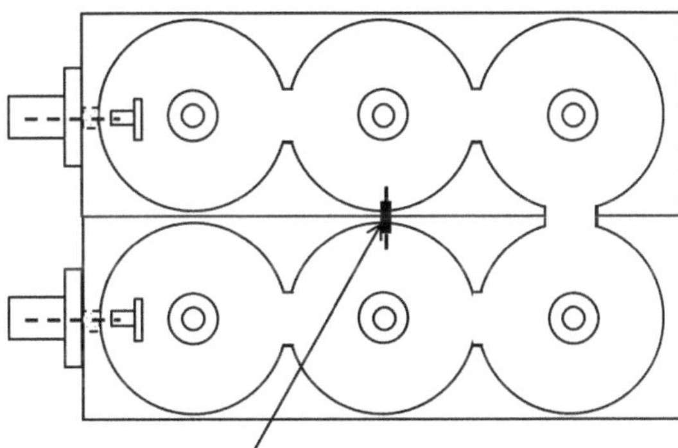

Fig. 8.1 Cross coupling between cavity-2 and cavity-5 in iris-coupled filter

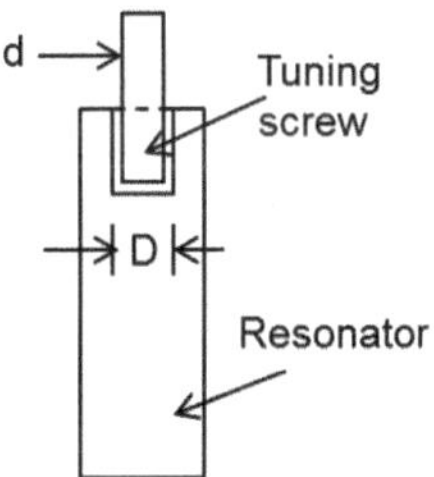

Fig. 8.2 Resonator—tuning screw sub-assembly

8.11 Duplexers

Duplexers are used to protect receivers from transmitters when using the same antenna for both transmission and reception. Duplexers are really duplex filters i.e. they have two filters designed to provide the required isolation between transmit and receive signal bands. A duplexer has two band pass filters, namely, TX and RX filter sections. Transmitter output is connected to the input of TX filter section. The output of RX filter section is connected to receiver input. The other end of TX and RX filter sections are coupled together to a common port, which is connected to antenna.

TX–RX isolation is specified in addition to the parameters specified for a band pass filter. A limit of 80 dB minimum for TX–RX isolation is quite common. The isolation limit is rejection characteristic for the design of TX and RX filter sections. Agilent application notes, AN 1287-8 and AN 1287-10 of Agilent could be referred for measuring TX–RX isolation of duplexers.

8.12 Developing Filter Design Software

Software packages in any high level language could be developed for the RF filter design tutorials presented in Chaps. 6 and 7. Excel tool could also be used for the tutorials. In addition to the computation of filter parameters, provision for updating the software based on internal filter design experience and external new developments in literature should be emphasised in the development of software. The developed software could be verified using the filter design tutorials and the examples available in the Microwave Filter Design Book [3]. Guidance is provided for the design of software for combline band pass filter and it could be extended for the design tutorials of other filters.

The sample electrical specification for the combline band pass filter, shown in Table 7.1 is reproduced below with internal limits only.

Electrical characteristics	Customer limit	Internal limit
Pass band centre frequency, fo	1 GHz	1 GHz
Bandwidth	80 MHz	85 MHz[a]
Insertion loss	2 dB max.	1.8 dB max.
Return loss	20 dB min.	22 dB min.
Rejection at (fo $\pm$ 100) MHz	60 dB min.	65 dB min.[b]
Pass band ripple (L_{ar})	0.4 dB max.	0.2 dB max.
Operating temperature	0–50 °C	0–50 °C
Input/output connectors	SMA (f)	SMA (f)

For calculating the combline band pass filter parameters:
[a] Recommended 80 MHz as bandwidth for design calculations. 85 MHz to be ensured while tuning the filter
[b] Recommended 70 dB as rejection at (fo $\pm$ 100) MHz for design calculations. 65 dB to be ensured while tuning the filter

> Key board inputs:
> f_o, Centre frequency (MHz): 1,000
> f_1, Lower band edge frequency (MHz): 960
> f_2, Upper band edge frequency (MHz): 1040
> f_u, Upper rejection frequency (MHz): 1100
> Rejection (dB) at upper rejection frequency, f_2: 70

L_{Ar} is the ripple in pass band. (t/b) is the ratio of the thickness of rectangular resonator to the width of cavity. Default value of 0.2 could be assumed for L_{Ar} and (t/b) in the software or it could be selected from a menu containing the standard values listed in the Microwave Filter Design Book [3] assuming second degree polynomial is available for all the Standard values of (t/b).

b, the width of cavity:

> Library of linked to the characteristics of filters developed over a period of time should be displayed for guidance. The required value of b in mm is entered.

With the software inputs listed above, the mechanical design parameters of the combline filter shown in Table 7.18 are displayed by the software except the blind hole details at the free end of resonators. Inputting the diameters of blind hole and tuning screw, the software displays the depth of the blind hole. Additional depth of 2 mm could be programmed in the software. The diameter for coupling screw is conveniently standardised to that of tuning screw.

References

1. HÖft M, Burger S, Magath T, Bartz O (2006) Compact combline filter with improved cross coupling assembly and temperature compensation. In: Proceedings of Asia-Pacific microwave conference
2. Thomas B (2003) Cross-coupling in coaxial cavity filters—a tutorial overview. IEEE Trans Microwave Theor Tech, 51(4): 1368–1376
3. Matthei GL, Young L, Jones EMT (1980) Microwave filters, impedance-matching networks, and coupling structures. Artech House, MA

Appendix A
Modes of Propagation

Electric and magnetic fields are associated with the propagation of electromagnetic waves in transmission lines. The modes of propagation indicate how the electric (E) and magnetic (H) fields are distributed relative to the direction of propagation in transmission lines. The electric and magnetic fields are always perpendicular to each other but they may or may not be perpendicular to the direction of propagation of electromagnetic waves. Three basic modes of propagation exist based on the distribution of E and H waves relative to the direction of propagation. The three modes of propagation are TEM, TE and TH modes.

The dominant mode propagation in coaxial lines and in homogeneous microstrip and stripline transmission lines is TEM. However, due to air gap or non-homogenous dielectric, higher order modes might exist. TM_{mn} and TE_{mn} higher order modes are associated with the transmission of electromagnetic waves in waveguides [1, 2].

TEM Mode

In Transverse Electro-Magnetic mode i.e. TEM mode of propagation, both electric and magnetic fields are perpendicular (transverse) to the direction of propagation of electromagnetic wave. Figure A.1 shows TEM mode in a RF transmission line.

TM Mode

In Transverse Magnetic mode i.e. TM mode of propagation, magnetic field is perpendicular to the direction of propagation of electromagnetic wave. The associated electric field is perpendicular to the magnetic field but it is at an angle, $\Phi°$, to the direction of propagation of electromagnetic wave as shown in Fig. A.2.

D. Natarajan, *A Practical Design of Lumped, Semi-Lumped and Microwave Cavity Filters*, Lecture Notes in Electrical Engineering,
DOI: 10.1007/978-3-642-32861-9, © Springer-Verlag Berlin Heidelberg 2013

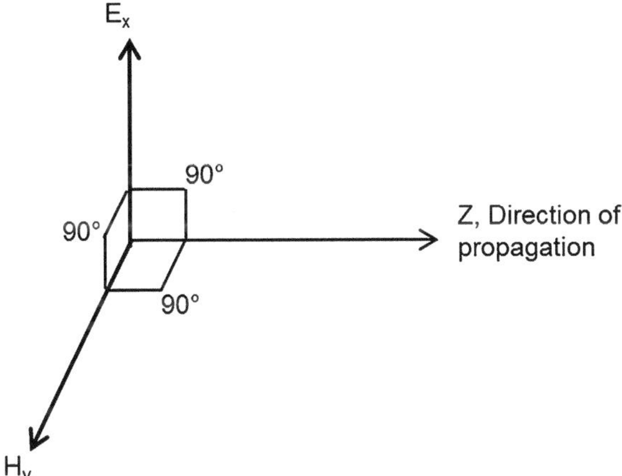

Fig. A.1 TEM mode of propagation

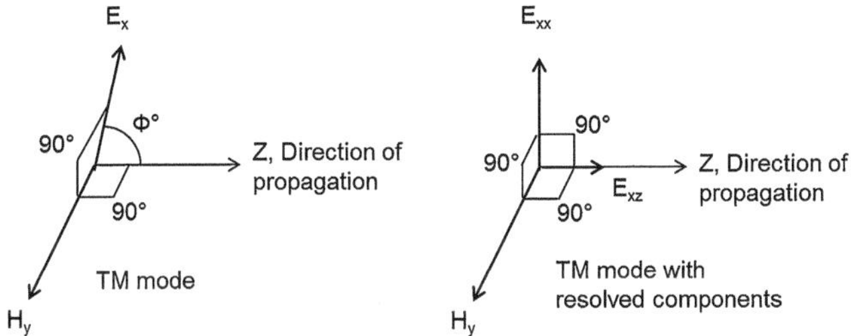

Fig. A.2 TM mode of propagation

The electric field, E_x, is resolved as E_{xx}, which is perpendicular to z, the direction of propagation and E_{xz}, which is in the direction of z. The resolved components are also shown in Fig. A.2. E_{xx} is called radial component and E_{xz} is called angular component of the electric field.

TE Mode

In Transverse Electric mode i.e. TE mode of propagation, electric field is perpendicular to the direction of propagation of electromagnetic wave. The associated magnetic field is perpendicular to the electric field but it is at an angle, $\Phi°$, to the direction of propagation of electromagnetic wave as shown in Fig. A.3.

The magnetic field, H_y, is resolved as H_{yy}, which is perpendicular to z, the direction of propagation, and H_{yz}, which is in the direction of z. The resolved components are also shown in Fig. A.3. H_{yy} is called radial component and H_{yz} is called angular component of the magnetic field.

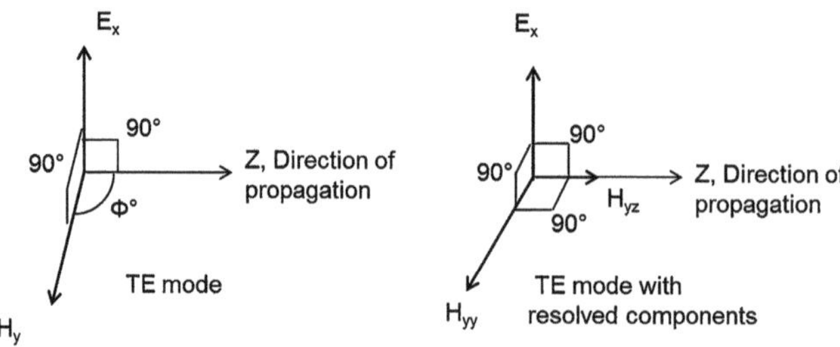

Fig. A.3 TE mode of propagation

References

1. Nghiem D, Williams JT, Jackson DR, Oliner AA (1995) Leakage of the dominant mode on a stripline with a small air gap. Microwave Theory Tech IEEE Trans 43(11):2549–2556
2. Landee RW, Davis DC, Albrecht AP (1957) Electronic designers' handbook. McGraw-Hill

Appendix B
Statistical Analysis on RF Filter Characteristics

The application of statistical analysis is quite popular in machining operations. It is extensively used for financial planning and market forecast. Statistical analysis could also be applied for designing and manufacturing RF filters. The application is explained for return loss characteristic in greater detail.

Normal Distribution For Return Loss

Let the customer specification limit for the return loss of RF filter is 20 dB min. The measured values of return loss on ten filters are shown in Table B.1.

For statistical analysis, 25 or more observations are required but it is limited to ten observations for explaining the application of statistical analysis. Mean and Standard deviation are statistical parameters and they are calculated for the measured values of return loss.

$$\text{Mean}, \mu = \sum_{i=1}^{n} x_i$$

Variable, x_i, is the measured values of return loss
n, the total number of measured values $= 10$
Mean, $\mu = 21.3$

$$\text{Standard deviation}, \sigma = \sqrt{\left[\sum_{i=1}^{n} \frac{(x_i - \mu)^2}{n-1} \right]}$$

Standard deviation, $\sigma = 0.5$

D. Natarajan, *A Practical Design of Lumped, Semi-Lumped and Microwave Cavity Filters*, Lecture Notes in Electrical Engineering, DOI: 10.1007/978-3-642-32861-9, © Springer-Verlag Berlin Heidelberg 2013

Table B.1 Measured values of return loss

Sl. No.	x_i, Return loss (dB)
1	20.7
2	21.4
3	21.5
4	20.7
5	22.0
6	21.9
7	20.8
8	22.0
9	21.0
10	21.1

In Table B.1, the return loss value, 20.7 dB is observed twice, the other values are observed once. The number of times a value is observed is called frequency. The frequency of observations is converted into cumulative distribution function (cdf) using statistical methods for calculating the percentage of RF filters with return loss less than 20 dB. The cdf, F (x), for the observed values of return loss from $(\mu - 3\sigma)$ to $(\mu + 3\sigma)$ is shown in Fig. B.1.

The cdf extends from $-\infty$ to $+\infty$ and the total area below the curve is unity. The specification limit, 20 dB, for the return loss is also shown in Fig. B.1. The area of the Normal curve to the right of 20 dB ordinate gives the percentage of filters that would be accepted and the area to the left of 20 dB ordinate gives the percentage of filters that would be rejected. The areas cannot be determined as a simple function of the variable, x, and hence the cdf is transformed to Standard Normal distribution [1] for determining the areas. The Standard Normal distribution has a mean of 0 and standard deviation of 1. The areas under Standard Normal curve are tabulated in statistical Tables for incremental values of ordinates and they could be easily read from the Table. The transformation of a variable, x, to Standard Normal curve is obtained by,

$$z = (x - \mu)/\sigma$$

Applying the transformation,

$z_{19.8}$
$$(19.8 - 21.3) / 0.5 = -3$$

$z_{20.0}$
$$(20.0 - 21.3) / 0.5 = -2.6$$

$z_{21.3}$
$$(21.3 - 21.3) / 0.5 = 0$$

$z_{22.8}$
$$(22.8 - 21.3) / 0.5 = +3$$

As the standard deviation is 1 for Standard Normal distribution, the return loss value, 19.8 dB, is termed as -3σ value and the return loss value, 22.8 dB, is termed

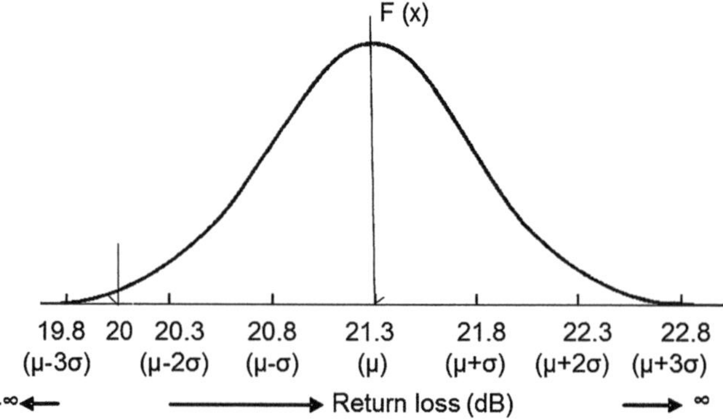

Fig. B.1 Normal (cdf) distribution for Return loss

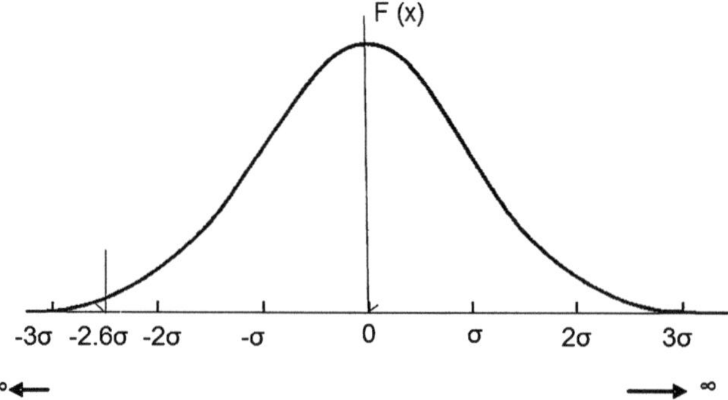

Fig. B.2 Standard Normal (cdf) distribution for Return loss, 20 dB min

as $+3\sigma$ value. The specification limit of 20 dB minimum corresponds to -2.6σ. The Standard Normal distribution for return loss is shown in Fig. B.2.

The shaded area to the left of -2.6σ in Fig. B.2 is equal to the shaded area to the left of 20 dB in Fig. B.1. Referring to the statistical Table for the area under Standard Normal curve, the area to the left of -2.6 is 0.005 i.e. 0.5 % of RF filters is expected to fail on return loss.

Setting Internal Specification For Return Loss

Assume the internal specification for return loss is set to 22 dB minimum. The measured values of return loss on ten filters are shown in Table B.2.

Mean, $\mu = 22.95$

Standard deviation, $\sigma = 0.58$

$\mu - 3\ \sigma$ $22.95 - 3 * 0.58 = 21.2$

$\mu + 3\ \sigma$ $22.95 + 3 * 0.58 = 24.7$

Table B.2 Measured values of return loss

Sl. No.	x_i, Return loss (dB)
1	22.2
2	22.5
3	23.0
4	22.4
5	23.3
6	23.0
7	23.7
8	24.0
9	23.1
10	22.3

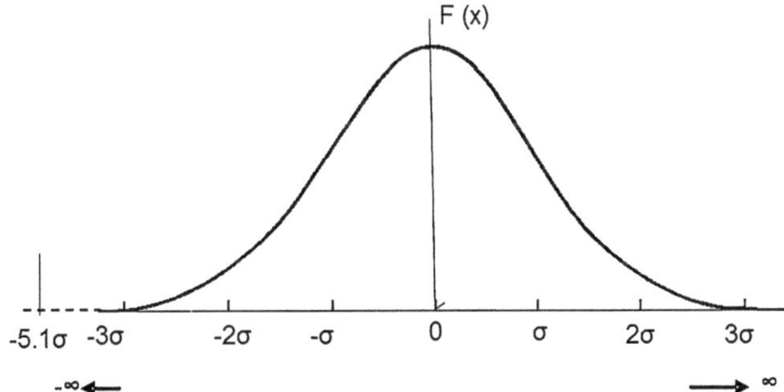

Fig. B.3 Standard Normal (cdf) distribution for Return loss, 22 dB min

Applying the transformation,

$z_{20.0}$ $(20.0 - 22.95) / 0.58 = -5.1$

$z_{21.2}$ $(21.2 - 22.95) / 0.58 = -3$

$z_{22.95}$ $(22.95 - 222.95) / 0.58 = 0$

$z_{24.7}$ $(24.7 - 22.95) / 0.58 = +3$

The Standard Normal distribution for return loss with internal specification limit of 22 dB minimum is shown in Fig. B.3.

With the internal specification limit of 22 dB minimum for return loss, the transformed 20 dB ordinate is -5.1σ in Standard Normal distribution. The area to the left of -5.1σ gives the percentage of rejection of RF filters in return loss. The area to the left of -5.1 is practically negligible and approximately 0.00002 % of RF filters is expected to fail on return loss. Even with inherent variations, the percentage of rejection is acceptable. Similar analysis can be made for all the characteristics of RF filters.

Reference

1. Christopher C (1983) Statistics for technology, a course for applied statistics. Chapman and Hall

Appendix C
Coaxial Connectors for RF Filters

RF coaxial connectors are available in many series. Some of the popular series of coaxial connectors are BNC, TNC, N and SMA. Connector size/power rating, frequency range, VSWR performance and cabling requirements are the major criteria for selecting the connector series. N and TNC series are available in two frequency ranges. General purpose N and TNC series are used up to 11 GHz. Precision N and TNC series are suitable for applications up to 18 GHz and VSWR characteristic is also specified.

Catalogues of manufactures on coaxial connectors indicate a wide range of plugs, sockets and adapters. Out of the types, the plugs, sockets and adapters that are useful for the design and manufacturing of RF filters are:

1. Coaxial plugs and sockets, panel mountable including field replaceable connectors
2. Cabling type plugs and sockets
3. In-series and between series coaxial adapters.

Coaxial Plugs and Sockets, Panel Mountable

Panel mountable coaxial plugs and sockets are used as input/output connectors for RF filters. Coaxial plugs and sockets are available without and with extended dielectric as shown in Fig. C.1. In connectors with extended dielectric, the PTFE insulator protrudes or extends beyond the mounting flange. The diameter of the PTFE insulator is designed for 50 Ω if a metal sleeve is sliding fitted over the extended dielectric.

Coaxial connectors are mounted on the walls of the cavity block of RF filters. If coaxial connectors without extended dielectric are used, there is discontinuity

D. Natarajan, *A Practical Design of Lumped, Semi-Lumped and Microwave Cavity Filters*, Lecture Notes in Electrical Engineering, DOI: 10.1007/978-3-642-32861-9, © Springer-Verlag Berlin Heidelberg 2013

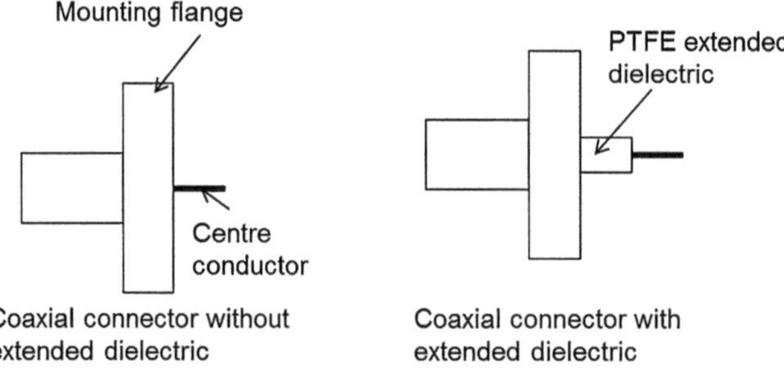

Fig. C.1 Types of coaxial connectors

(mismatch) in 50 Ω for the length equal to the wall thickness of cavity block. The discontinuity results in mismatch in the tuning of microwave cavity filters. Hence, coaxial connectors with extended dielectric are used in RF filters. The length of the extended dielectric is equal to the wall thickness of cavity block. The walls of the cavity block over the extended dielectric ensure continuity in 50 Ω impedance. The diameters of the holes in the walls of cavity block are dimensioned for sliding fit over the extended dielectric of coaxial connectors. Panel mountable coaxial plugs and sockets with extended dielectric are available in all coaxial connector series.

Feed Through Terminals

Feed through terminals are used in place of input/output coaxial connectors in surface mounted filters. Feed through terminals are designed with brass male centre contact, which is force fitted into PTFE insulator. The diameters of the centre contact and PTFE insulator are designed for 50 Ω impedance (see Sect. 2.5). The captivation of centre contact is ensured in the design of feed through terminal. The length of PTFE insulator is matched to the wall thickness of cavity block. The length and diameter of centre contact which protrudes outside the filter for the surface mounting of filter is matched to the interface dimensions of field replaceable coaxial connectors. Field replaceable connectors are mated temporarily with the feed through terminals for tuning and inspection of filters.

Cabling Type Plugs and Sockets

RF filters are connected to VNA (Vector Network Analyser) for tuning and demonstrating compliance to specifications. The device under test (DUT) is connected to the VNA ports using the standard cables supplied along with VNA. The standard cables are available in different lengths and are characterised by low VSWR (high return loss) and high stability. They are also mechanically flexible for connecting DUTs. The standard cables are available with precision types of

Table C.1 Examples of in-series and between series adapters

In-series adapter	Between series adapter
BNC plug–BNC socket	7 mm–BNC plug
TNC plug–TNC plug	N socket–TNC socket
N socket – N socket	SMA plug–BNC socket
SMA socket–SMA plug	TNC plug–SMA socket

coaxial connectors such as 7 mm or N or SMA3.5. The standard cables are used for tuning or testing high frequency RF filters.

For tuning and testing RF filters with BNC or TNC input/output connectors, test cables are assembled using cabling type plugs and sockets. The return loss of the assembled test cables is tested before use to ensure that it is better than 30 dB in the frequency band of interest. For better return loss and stability, cabling plugs or sockets with captivated centre contacts are preferred. Coaxial cables such as RG142B/u and RG214/u are recommended for the assembly of test cables. Cable assembly procedure recommended by the manufacturers of coaxial connectors should be followed to ensure better return loss for test cables.

In-Series and Between Series Coaxial Adapters

Coaxial adapters are used to interconnect two coaxial connectors which may plugs or sockets. If a coaxial adapter interconnects two coaxial connectors of same series, the adapter is known as in-series adapter. If a coaxial adapter interconnects two coaxial connectors of different series, the adapter is known as between-series adapter. Examples if the adapters are shown in Table C.1. General purpose and precision adapters are available. VSWR is specified for precision adapters.

Coaxial adapters are required for the tuning and testing of RF filters. They are also required for the initial set-up calibration of VNA. Precision adapters are recommended for all RF measurements. Precision in-series and between series adapters are available with SMA3.5, TNC-18 GHz, N-18 GHz and 7 mm coaxial connectors. Torque spanners recommended by the manufacturers of coaxial connectors should be used for tightening the coupling nuts of coaxial connectors to ensure good RF contact and to prevent damage to the adapters.

About the Author

Dhanasekharan Natarajan, Electronics engineer from College of Engineering, Guindy, Chennai in 1970, obtained his post-graduate in Engineering Production (Q&R Option) from the University of Birmingham, UK in1984. He is a Senior Member (00064352-member for 25 years), IEEE (MTT Soc.) and his biography is published by Marquis, USA in their fourth edition, "Who's Who in Science and Engineering".

He retired as Asst. Professor in RV College of Engineering, Bangalore. His earlier professional achievements at Bharat Electronics and Radiall Protectron include Application of modern Q&R techniques for defence equipments, Designing and implementing computerized Quality Management System, Designing software for Optical interferometer and Design/manufacturing of lumped, semi-lumped and microwave cavity filters using self-developed software.

D. Natarajan, *A Practical Design of Lumped, Semi-Lumped and Microwave Cavity Filters*, Lecture Notes in Electrical Engineering,
DOI: 10.1007/978-3-642-32861-9, © Springer-Verlag Berlin Heidelberg 2013

Index

D. Natarajan, *A Practical Design of Lumped, Semi-Lumped and Microwave Cavity Filters*, Lecture Notes in Electrical Engineering,
DOI: 10.1007/978-3-642-32861-9, © Springer-Verlag Berlin Heidelberg 2013